Biomass for Energy in the Developing Countries

Other Pergamon Titles of Interest

AUER	Energy and the Developing Countries
BRAUNSTEIN	Biomass Energy Systems and the Environment
EGGERS LURA	Solar Energy in Developing Countries
FAZZOLARE	Beyond the Energy Crisis (5 volumes)
FERNANDES	Building Energy Management
FRITZ	Future Energy Consumption of the Third World
GARDEL	Energy: Economy and Prospective
HELCKE	The Energy Saving Guide
MCVEIGH	Sun Power, 2nd Edition
O'CALLAGHAN	Design and Management for Energy Conservation
PALZ & STEEMERS	Solar Houses in Europe: How They have Worked
ROSS	Energy from the Waves, 2nd Edition
TAHER	Energy: A Global Outlook
TWIDELL	Energy for Rural and Island Communities 1981
VOGT	Energy Conservation and Use of Renewable Energies in the Bio-industries

Pergamon Related Journals *(Free specimen copy gladly sent on request)*

Energy
Energy Conversion and Management
International Journal of Hydrogen Energy
Journal of Heat Recovery Systems
OPEC Review
Progress in Energy and Combustion Science
Solar Energy
Sun World

Biomass for Energy in the Developing Countries

CURRENT ROLE, POTENTIAL, PROBLEMS, PROSPECTS

by

D. O. HALL
Department of Plant Sciences
University of London, King's College

G. W. BARNARD
Department of Chemical Engineering
Imperial College, London

and

P. A. MOSS
Department of Plant Sciences
University of London, King's College

PERGAMON PRESS

OXFORD · NEW YORK · TORONTO · SYDNEY · PARIS · FRANKFURT

U.K.	Pergamon Press Ltd., Headington Hill Hall, Oxford OX3 0BW, England
U.S.A.	Pergamon Press Inc., Maxwell House, Fairview Park, Elmsford, New York 10523, U.S.A.
CANADA	Pergamon Press Canada Ltd., Suite 104, 150 Consumers Rd., Willowdale, Ontario M2J 1P9, Canada
AUSTRALIA	Pergamon Press (Aust.) Pty. Ltd., P.O. Box 544, Potts Point, N.S.W. 2011, Australia
FRANCE	Pergamon Press SARL, 24 rue des Ecoles, 75240 Paris, Cedex 05, France
FEDERAL REPUBLIC OF GERMANY	Pergamon Press GmbH, 6242 Kronberg-Taunus, Hammerweg 6, Federal Republic of Germany

First edition 1982

Library of Congress Cataloging in Publication Data

Hall, D. O. (David Oakley)
Biomass for energy in the developing countries
Includes bibliographical references.
1. Biomass energy. I. Barnard, G. W. II. Moss, P. A. III. Title.
TP360.H35 1982 333.95'3 82-322
AACR2

British Library Cataloguing in Publication Data

Hall, D.O.
Biomass for energy in the developing countries.
1. Biomass energy - Underdeveloped areas
I. Title II. Barnard, G.W. III. Moss, P.A.
622'.6 T360

ISBN 0-08-029313-1 (Hardcover)
ISBN 0-08-028689-5 (Flexicover)

In order to make this volume available as economically and as rapidly as possible the author's typescript has been reproduced in its original form. This method unfortunately has its typographical limitations but it is hoped that they in no way distract the reader.

Printed in Great Britain by A. Wheaton & Co. Ltd., Exeter

Preface

The authors of this book have been actively involved in Biomass Energy research for several years and have worked with a variety of national and international agencies, including the United Nations and the Solar Energy Research Institute. So as to keep the costs of publication to a minimum, the authors have agreed to forego any potential royalties from the sale of this book. They welcome any comments and corrections with regard to the text and are particularly anxious to receive additional information on current Biomass Energy use in Developing Countries so that their data base can be continually improved.

Contents

Summary and Conclusions

I General Considerations

1. The importance of biomass energy in the developing countries today is indisputable. In many countries, biomass supplies more than half of the total amount of energy used and in some this reliance is as great as 95%.

2. These startlingly high figures result from the fact that the traditional biomass fuels - wood, charcoal, crop residues and dung - are still the primary cooking and heating fuels for the vast majority of the population of the developing countries, especially the rural and urban poor.

3. The role of biomass energy is in jeopardy, however. In many countries, biomass fuel supplies can no longer meet demand and for a variety of reasons, severe shortages are developing.

4. These problems, sometimes known as the 'second energy crisis', are superimposed on those caused by the rising price of oil. Despite their modest consumption levels, many of the developing countries have been hard hit by the oil crisis and are now facing acute balance of payments difficulties as a consequence.

5. Because of these dual energy problems, there is a clear need to develop indigenous energy sources in the developing countries - renewable as well as non-renewable. This is particularly urgent in view of the fact that expanding energy supplies will be necessary simply to maintain present levels of consumption for a growing population - let alone providing for the major increase in energy consumption that will be essential if economic development is to proceed.

6. Amongst the various options, there is significant scope for the development of biomass energy systems. Present biomass resources can be utilised more efficiently and a variety of approaches, new and old, are available for growing and using biomass for energy.

I General Considerations (contd)

7. For supplying commercial energy needs, a range of processes exist for producing convenient liquid, gaseous and solid fuels from biomass. Although they will not necessarily be cheap, in principle these could allow biomass energy to replace oil in each of its present roles. For individual countries, these options must be assessed in the light of local conditions and the availability of alternative energy sources - fossil fuels, hydro-power, etc.

8. For supplying domestic energy needs, on the other hand, there are very few viable alternatives to biomass. For the vast majority of the poorer sectors of the developing countries, locally grown biomass fuels are the only source of energy which is within their financial reach. This situation is unlikely to change in the near future since, without heavy subsidies, virtually all non-biomass alternatives - kerosene, bottled gas, etc. - are much too expensive to have any widespread impact, other than perhaps in towns. Developing low-cost methods of producing and using biomass energy is therefore of the highest priority if the future domestic energy requirements of the poor are to be met.

9. The biomass energy systems that are described in this book cover the whole spectrum of possible applications of biomass energy. The technologies involved form a large and heterogeneous group - ranging from simple wood stoves to large and highly sophisticated processes for converting biomass to liquid fuels. In this book, biomass systems are discussed under the headings : Energy for Domestic Needs, Energy for Rural Development, and Energy for Petroleum Replacement.

10. Because of the heterogeneous nature of biomass energy systems it is difficult to make generalisations about their properties or their potential. One feature that they do have in common, however, is the high degree to which they are interlinked with other factors - environmental, agricultural, socio-economic, and political.

11. It is vital that these interactions are understood from the outset since they are of crucial importance in determining the limitations on biomass energy systems, as well as their potential impact both good and bad.

12. Limitations on the use of biomass energy systems, for example, can occur for a number of technical and economic reasons. One of the most important of these constraints is the competition for land between food and fuel production. In many cases, the introduction of new biomass systems will also be restricted by lack of financial and other resources, and the absence of the necessary infrastructures, extension services, manufacturing capabilities and technical know-how.

13. Barriers to the implementation of biomass systems can also arise for a variety of other reasons, more related to social

I General Considerations (contd)

and political factors. Conflicts of interest can be a major problem, for example, in cases where present ownership and control over resources is unequal, and where the introduction of new systems is likely to benefit some groups more than others. Other factors which can slow down the rate of diffusion of new energy schemes include lack of perception of energy problems and general unwillingness to invest in systems which have not been fully proven.

14. One of the advantages of biomass energy systems, however, is that they often have the potential for yielding subsidiary benefits other than their simple effect on energy supplies. Some systems, for example, can help to improve the environment whilst others can provide a stimulus for rural development through offering a source of employment, income and energy in rural areas.

15. It is important to recognise, on the other hand, that none of these benefits is necessarily guaranteed. As with most other technologies, if they are applied without consideration of their possible environmental hazards, and without taking into account the needs of the local population, biomass systems can equally well have a negative impact - both in damaging the environment and in increasing the exploitation of rural areas and rural populations.

16. Assessing the potential of biomass energy in the developing countries and arriving at effective implementation policies, is undoubtedly a very complex task. There are many limitations and constraints to take into account, and many decisions to be made concerning the choice of systems, scale options and the implementation approaches to be taken.

17. Furthermore, there are a large number of unknowns involved in biomass energy development. Whilst some biomass systems have been tried and found to be successful under some conditions, others have yet to be tested on a large scale. In almost all cases, detailed local surveys and field testing will be essential since many of the relevant variables - social, economic and technical - are highly site-specific and because very little information generally exists with regard to current patterns of biomass energy use. Adaptation of systems to fit local conditions is also likely to play a crucial part in ensuring the success of biomass energy schemes.

18. Overall, biomass energy will not be a simple solution to the energy problems of the developing countries. Biomass systems will not necessarily be cheap, nor will they be implemented easily, without a major commitment from governments and a considerable amount of political will at the national and international level.

I <u>General Considerations (contd)</u>

19. The era of cheap oil and abundant traditional fuels is over, however, for practically all developing countries. Coming to grips with the fundamental problems that energy supply poses, and arriving at effective solutions, is one of the most vital tasks in the overall development process. In view of the urgency of the problems and the lack of viable alternatives, it is clear that biomass systems have an important role to play.

II Energy for Domestic Needs

A Nature and Cause of Problems

20. Many millions of people in the developing countries are now facing acute shortages of traditional biomass fuels. These shortages carry with them immense costs, both in human and environmental terms, although the economic costs are often indirect and hard to quantify.

21. The situation elsewhere in the developing countries, whilst not as serious yet, is rapidly deteriorating due to the combined effects of growing demand for traditional fuels and diminishing supplies.

22. The increase in demand for these fuels is closely related to population growth, but it is accentuated by urbanisation - which concentrates demands - and by oil price rises, which put alternative fuels out of the financial reach of the poor.

23. Whilst deforestation is certainly the major reason for the decrease in fuel supplies, fuelwood cutting is not the only cause of deforestation. Other factors that are often equally important in causing deforestation include the clearing of land for agricultural purposes, overgrazing by animals and uncontrolled logging. It is vital that the true causes of deforestation are appreciated if effective counter-measures are to be devised.

24. As well as fuel shortages existing in absolute terms, it should also be stressed that the hardship that they cause can be greatly increased through distributional inequalities. In cases where access to resources is in the hands of the wealthy, along with the ability to pay for alternatives, it is the poor who suffer most from fuel shortages.

B Possible Technical Solutions

25. To help reduce domestic energy problems in the developing countries a variety of approaches is possible. These centre around either increasing the production of biomass fuels or improving the efficiency of fuel processing and fuel use. The approaches that appear to have the greatest immediate potential are :

a) Better use of forest resources

In many countries, the simplest way of expanding fuelwood supplies is to increase the efficiency of utilisation of existing forest resources. This can include, for example, making fuller use of logging wastes, opening up new areas of forest for fuelwood extraction, relaxing legal constraints on forest use and promoting replanting. In each case it is

II Energy for Domestic Needs (contd)

essential that over-cutting is avoided so as to ensure the renewability of the resource.

b) Planting fast growing trees on unused land

This can be done on a large scale, to supply urban needs, or on a smaller scale for village/individual uses. It can be organised commercially, through forestry organisations, or on a communal or private basis. Where land is scarce, Agroforestry techniques can be useful since they permit food and fuel to be produced on the same piece of land. Planting multiple-use tree species can also help the economics of tree growing considerably. Leguminous trees have the added advantage of improving soil fertility.

c) Rationalisation of charcoal-making activities

Present methods of charcoal making are often extremely inefficient. Reasonably cheap medium-sized masonry or brick kilns can be used, however, which are twice as efficient as traditional methods. These can be profitably employed in commercial or community charcoal making operations where the necessary capital is available. Efforts to prevent over-cutting and promote replanting will also help to reduce the rate of deforestation caused by charcoal making and increase the renewability of charcoal supplies. The potential for introducing more efficient kilns to small-scale charcoal makers is not great, due to cost constraints.

d) Introduction of improved cooking stoves

Using locally available skills and materials, low cost cooking stoves can be made which are substantially more efficient than traditional stoves or open fires. If this increased efficiency could be translated into practice, under field conditions, it could lead to considerable fuel savings. For stoves to be accepted by users, however, they must be designed so that they match local cooking habits and are convenient to use with the available fuels. They must also fulfil the subsidiary requirements of present stoves and fires, such as room heating and keeping away insects. It should be noted that actual fuel savings may not be as great as predicted from laboratory tests since users may prefer to burn the same amount of fuel and simply use more heat.

e) Biogas

The production of biogas from animal dung and other organic wastes has a number of theoretical advantages in that it produces a fuel that is well suited for cooking and lighting purposes whilst retaining nutrients and improving hygiene. The use of household biogas plants will probably be limited in many cases, however, by the relatively high costs of the plants and the shortage of suitable feedstock materials, particularly for the poor. Cultural considerations are also important in determining the acceptability of using dung and human wastes for fuel. To reduce capital costs and increase the scope of biogas technology so that poorer members of the community can contribute - and

II Energy for Domestic Needs (contd)

benefit - larger scale communal plants have some potential. These will require a substantial degree of village cooperation, however, if organisational and distributional problems are to be avoided.

C Socio-economic and Political Considerations

26. The technical feasibility of the various household energy schemes mentioned above will vary tremendously from location to location, depending on factors such as land-use practices, agro-climatic conditions, population density and competing uses for biomass resources.

27. Equally important, however, are the many socio-economic and political factors which impinge upon present fuel-use practices - and which will have a direct effect on the rate of diffusion of new energy schemes and their subsequent impact.

28. Some of the most important of these considerations are listed as follows :

- The level of perception of energy problems and the importance of traditions in fuel collection and fuel use practices - and the logic behind them
- Poverty and the general shortage of land, capital and other resources necessary for investment in new energy systems
- Competition for use of resources with other activities
- Constraints on the choices open to individuals due to the importance of short-term considerations, and the avoidance of risk, as opposed to long-term needs
- Unequal land distribution at the village level and inequalities in the ownership of resources, access to credit and general political and economic power
- Conflicts of interest between members of the community over the distribution of costs and benefits from new energy schemes
- The value placed on non-financial benefits from new energy systems
- The involvement of women in the decision making process

II Energy for Domestic Needs (contd)

D Requirements for Successful Implementation

29. Given the complexity of many of these issues, it is vital that the perceptions of potential users - and the problems and constraints facing them - are fully understood in attempting to devise effective domestic energy strategies. This understanding can often only come through a close interaction with the people involved.

30. For this reason, a 'bottom up' approach to the diffusion of new energy systems is generally much more likely to succeed than a 'top down' approach in which people's needs are assessed from outside and then efforts are made to design and diffuse 'appropriate' energy solutions to solve their problems.

31. A 'bottom-up' approach is not easy to achieve in practice, however. It involves organising and assisting people to solve their own problems - by defining their own energy needs, developing locally acceptable responses (technical and organisational), and carrying out mutually agreed programmes.

32. Cooperation is obviously vital, but for this to be forthcoming people must view energy schemes as being for their benefit. Developing confidence in government extension services is a crucial part of this, and a combination of persuasion, demonstration and suitable incentives may often be required to bring about the necessary transition in people's attitudes. Employing women extension officers can be helpful in this context, in encouraging the participation of women in developing and carrying out energy programmes.

33. Building up effective extension services will be a problem in almost all developing countries. It is, however, an essential part of the process of developing and diffusing domestic energy schemes, since market forces have generally failed to produce viable long-term solutions to the energy problems of the poor.

34. In summary, the domestic energy problems of the developing countries are often extremely intractable. Present energy-use practices are an integral part of everyday life and, as such, are closely tied with the basic problems of poverty and inequality. These practices will not change overnight, nor will the problems disappear without a considerable degree of commitment, imagination and action from national governments. These matters must be given highest priority, however, since if domestic energy problems are not tackled now the price to pay later both in economic terms, and in human hardship, is sure to increase.

III Energy for Rural Development

A Background

35. Providing a sustainable and affordable energy source is an important component of rural development. Coupled with other inputs - financial, technological, etc. - a relatively small increase in energy use can bring with it substantial improvements in agricultural productivity and can permit growth in rural industries and public services.

36. Biomass energy systems have particular potential in this context, due to their versatility and suitability for decentralised rural applications. Simple and relatively low-cost conversion devices are available that can supply all forms of rural energy demands, providing that a suitable source of biomass is available and given that moderate levels of mechanical skills exist for operation and maintenance.

37. Costs, in some cases, are already competitive with conventional commercial fuels and as oil prices rise biomass systems should become increasingly attractive.

B Technical Options

38. Biomass fuels - wood, charcoal, sawmill wastes, etc. - are already used to a substantial degree for providing heat for rural industries. This use could be expanded considerably providing that suitable biomass sources exist.

39. For supplying mechanical energy needs the approaches that appear to have the greatest immediate potential are :

a) Gasifiers

Small-scale gasifiers can be built using simple materials, enabling dry forms of biomass (wood chips, rice husks, corn cobs etc.) to be converted to a low energy gas suitable for use in diesel engines - mobile as well as stationary. Diesel consumption can be reduced by up to 80% but careful maintenance is required to ensure smooth operation.

b) Biogas

Animal dung, various organic wastes, and certain types of degradeable plant matter can be converted to biogas (a methane/carbon dioxide mixture) through anaerobic digestion. This can also be used as a supplementary fuel for diesel engines although its use will be generally limited to stationary power applications due to the problems of gas storage.

c) Steam Engines

This is a well-tried alternative for using wood for medium-large scale mechanical power applications which could be revived.

d) Liquid Fuels

The simplest approach to producing liquid fuels on a small-scale local level is to grow crops such as oil palm, coconut, sunflower and sesame from which vegetable oils can be extracted. These can be used directly as a diesel extender and with relatively minor treatment can be used indiluted. There is also considerable potential for the small-scale production of ethanol from sugar or starch crops although there has been little experience in this area as yet.

e) Draught Animals

These are the traditional method of supplying mechanical power in rural areas. They have many advantages and the potential that exists for increasing their use should not be ignored.

40. The most important technical limitation on the use of these systems arises in the supply of adequate quantities of biomass materials. Some of the key considerations in this context are as follows :

- Competition for land use
- Competition between using biomass for food, fuel or other purposes
- Competition between commercial and domestic uses of biomass fuels.

41. New biomass systems are unlikely to be accepted if they are unprofitable, but it is vitally important that expanding the use of biomass for new rural applications does not have the effect of increasing domestic fuel shortages, or creating further environmental damage.

42. Possible options for increasing biomass supplies include the following :

- Increasing the use of waste products from agriculture and rural industries

III Energy for Rural Development (contd)

- Growing trees or other energy crops on unused land
- Use of improved agricultural techniques so as to increase both food and fuel production
- Use of Agroforestry techniques.

43. In almost all areas, a significant increase in biomass production is theoretically feasible and in principle this could allow a substantial degree of energy self-sufficiency in rural areas to be attained. Energy is not the only input that is required, however, and in practice it will often be the lack of financial resources, technical know-how and other essential inputs which limit the impact of biomass energy systems.

C Social Considerations

44. Organisational inputs can also be important, since encouraging individuals and communities to work together in pooling resources and skills is often a prerequisite for maximising production efficiency. Where conflicts of interests are strong, achieving this mutual cooperation may be very difficult.

45. It should also be recognised that increasing the productivity of rural areas through the expansion of energy supplies will not in itself guarantee an equitable distribution of the benefits. If new energy devices are used to displace labour, for example, this can create severe problems for poorer sectors of the community who often rely on hiring out their labour to earn a living.

D Implementation

46. Most of the uses of biomass energy mentioned in this category have the potential of yielding a direct financial benefit to the user either in the form of increased agricultural or industrial productivity, or as decreased energy bills. It is likely, therefore, that market forces will be more effective in diffusing these technologies than for domestic energy systems - which only yield indirect benefits such as increased convenience.

47. National governments have an important role to play, however, introducing these biomass systems. Some of the more crucial government roles include :

- Provision of credit facilities and other incentives
- Performing research and development work

III <u>Energy for Rural Development (contd)</u>

- Providing demonstrations and technical assistance
- Monitoring biomass collection practices to prevent over-cutting of forests,etc.
- Adjusting pricing and taxation levels for commercial fuels so as to support biomass promotion strategies

IV Energy for Petroleum Replacement

A Background

48. The problems caused by rising oil prices have provided very strong incentives for developing countries to develop alternative energy sources. Using biomass energy for petroleum replacement is a relatively new concept which may have considerable potential, especially in countries where large areas of unused land are available.

49. The main uses of petroleum energy in the developing countries are for transportation fuels, industrial heat and electricity generation. To use biomass energy in these applications, biomass must first be converted to a form in which it can substitute for petroleum products, as well as being transported to the areas (usually urban) where demands exist. Both of these steps will involve costs in financial terms as well as in energy.

B Industrial Heat and Electricity Generation

50. For industrial heat and electricity generation, wood is generally the most economic biomass fuel. Techniques for burning wood are well advanced and combustion efficiencies can be almost as high as for oil burning facilities if well designed equipment is used.

51. In cases where existing oil-burning facilities are being converted to operate on wood, gasification of the wood as a preliminary step can be a useful approach. The low energy gas produced can be burnt in gas turbines, diesel engines, and oil-fired boilers and furnaces, with only minor equipment modifications. This can cut down conversion costs as well as increasing control over the process.

52. The most efficient method of producing wood is to use short-rotation, fast-growing tree plantaions. Under good conditions, annual wood yields of up to 15-20 tonnes per hectare are possible at costs that are below the equivalent energy cost of oil. Using improved tree varieties, even higher yields may be feasible.

53. Because of the high costs of transporting wood, the most economic method of using wood energy is to locate industries and electricity generating facilities near to tree plantations.

54. In cases where this is not possible, conversion of wood to charcoal can improve the economics of transportation considerably - since charcoal has roughly twice the energy content of wood on a weight basis. Charcoal is also a more efficient fuel to burn and has a variety of additional properties that make it useful for steel-making and cement manufacture. Its primary disadvantage is the fact that a significant amount of the energy content of the wood (at least 50%) is lost during the conversion process.

C Transportation Fuels

55. A wide variety of theoretical approaches exist for using biomass energy to replace gasoline and diesel as transportation fuels. Of these, the techniques which are available now, or will be in the relatively near future, are :

a) Use of small mobile gasifiers coupled to modified diesel or gasoline engines

b) Production of ethanol by fermentation of sugar crops, starch crops or surplus molasses

c) Production of methanol from wood - through gasification to produce synthesis gas followed by catalytic conversion to methanol

d) Production of gasoline from wood using gasification, as above, followed by Fischer-Tropsch conversion to gasoline. Alternatively wood-derived methanol can be converted to gasoline using the 'Mobil process'

e) Production of vegetable oils from crops such as oil palm, sunflower, coconut palm, groundnut, sesame and soya beans

56. Of these options the production of ethanol from sugar cane is the only approach that is presently in use on a significant scale in the developing countries. There are, therefore, major uncertainties regarding their suitability, both in technical and economic terms. Many of the relevant economic parameters (biomass production costs, transportation costs, labour and capital costs, etc.) are also very site-specific, so economic appraisals will require detailed local surveys. Whilst none of these systems are likely to be cheap, they offer one of the few liquid fuel alternatives for countries without fossil fuel reserves - and the only one available at present which is renewable.

IV Energy for Petroleum Replacement (contd)

D System Choice and Implementation Strategy

57. The choices open to developing countries for replacing petroleum energy, and the relative merits of the various approaches mentioned, will obviously depend on a wide variety of local considerations. Some of these are listed as follows :

- The amount of unused land that is available and its suitability for growing energy crops
- The location of available land with respect to centres of demand and the resulting transportation problems
- Competition between using land for energy purposes as opposed to food production
- The availability of capital, foreign exchange, trained manpower and other necessary inputs
- Economies of scale
- The importance of environmental considerations
- The alternative energy sources available.

58. It should be noted, however, that for many of these technologies there is considerable scope for adapting systems to suit local conditions. This might involve, for example, developing simplified and smaller-scale plant designs and using labour-intensive techniques for growing biomass. Building up indigenous R and D capabilities will be an essential part of this process. The assistance of the developed nations will also be crucial, both in providing financial and technical backing and in facilitating free and unrestricted transfer of technologies.

59. There are also a number of important social and developmental considerations to be taken into account in designing petroleum replacement systems. These include fundamental questions such as who will benefit from new systems? Will the profits be recycled to stimulate rural development or will the net result be that biomass energy systems are used as a way of satisfying the energy needs of the affluent urban sectors at the expense of the rural poor?

60. It is clear, therefore, that as well as the many technical and economic questions that must be answered in assessing the feasibility of biomass energy systems, there is also a range of broader political issues at stake. How these issues are resolved will have a major impact in determining whether the full potential of biomass energy - in developmental terms - is realised.

Recommendations

I National Level

1. The growing domestic fuel shortages in many developing countries and the worsening effects of the oil crisis underline the urgent need for governments to develop effective energy strategies to combat these problems. Formulating policies towards biomass energy must be an integral part of this process.

2. Governments are faced with a difficult task in devising effective biomass energy policies. Not only are there many technical and economic considerations to take into account, there is also a whole range of social, political and environmental issues to address as part of the planning process. Understanding these wider issues is a crucial part of arriving at realistic energy policies since, in many instances, non-technical considerations will be the primary determinants of success or failure in energy projects.

3. In view of the complexity of many of the issues involved in developing biomass energy policies, and their highly site-specific nature, it is not possible at this stage to make any specific recommendations regarding which policy approaches should be taken or which biomass systems should be promoted. Countries must formulate policies on an individual basis - in the light of their own needs and their own development priorities, and in the context of their own socio-political and economic circumstances.

4. It is relevant, however, to point out some of the more critical elements of the policy making process, particularly those aspects which have tended to be overlooked in the past. Countries should pay careful attention to each of the following :

   a) Recognising the role that biomass energy already plays in national energy supplies. Detailed preliminary surveys must be performed to establish the patterns of use of biomass fuel and to determine the nature and underlying cause of domestic energy problems.

   b) Understanding the requirements and demands of present and potential biomass users so that systems can be designed which are affordable, as well as being attractive to the individual and appropriate to their energy needs. This will require detailed knowledge of the perceptions and attitudes of

I National Level (contd)

individual users and a full appreciation of the constraints that they face in making energy decisions.

c) Paying specific attention to the energy needs of the poor, and of rural areas, and the special problems that they face in supplying the financial and other inputs that are needed to adopt new energy systems. Governments must place emphasis on developing low-cost systems, and must take measures to target necessary assistance to those whose energy problems are most severe.

d) Assessing biomass technologies in a critical and realistic light. Careful testing and evaluation of systems under field conditions is essential since many biomass systems have yet to be fully proven - despite the claims of salesmen and enthusiasts. Adaptation of systems to suit local requirements and conditions is also likely to be a critical part of assuring the success of biomass energy schemes.

e) Recognising and confronting the conflicts of interest that arise in biomass energy development. These include conflicts between rural and urban needs, conflicts between individuals or groups within the community, conflicts between food and fuel production, and conflicts between short-term and long-term priorities. Policies which fail to account for and resolve these conflicts are unlikely either to succeed or to have the desired impacts.

5. To address adequately these many issues, and to be able to evaluate technologies and arrive at effective policy decisions, governments will require strong indigenous planning capacities. Governments should also place a high priority on developing local research capabilities since, as well as their role in adaptation and development of technologies, research centres can play a valuable part in serving as a focal point for training personnel and demonstrating new techniques.

6. The strengthening of extension services will be a very important prerequisite for any large-scale implementation efforts. Coordinating biomass energy schemes with other rural development activities is essential if the maximum use is to be made of resources and manpower. Developing trust and effective communication between extension officers and user groups will also be vital if their needs are to be accurately perceived and their problems understood. Effective communications between field officers and central planning and research facilities will be equally important if these needs are to be transmitted and incorporated into policy decisions.

I National Level (contd)

7. Developing an effective planning, research and implementation machinery will be a major challenge for practically all developing countries. Apart from the general problems relating to shortages of resources and manpower, there are often basic problems to deal with such as lack of cooperation between government agencies, poor relationships between extension officers and members of the community, and lack of motivation. The low prestige associated with jobs relating to rural development activities is also a common problem which makes it hard to attract qualified and dedicated personnel to crucial posts in agriculture, forestry and other ministries. Governments must tackle these problems if they are ever to get on top of the energy and rural development questions at stake.

8. It is important that governments approach biomass energy development in an integrated fashion. Individual elements of energy planning, such as fuel pricing policies and subsidies, must be designed so they are complementary rather than contradictory. At a higher level, energy planning must be fully integrated with overall development planning - since energy problems are so closely linked with many of the other problems facing developing countries that it will rarely be possible to solve them effectively in isolation.

II International Level

9. Financial and technical assistance from the industrialised nations and from the international agencies will be crucial in helping developing countries to formulate and carry out biomass energy programmes. Action should be taken now to provide appropriate mechanisms for this to occur.

10. In the short term, many of the requirements will be for relatively small-scale financial inputs and for technical expertise. Key areas will be :

a) Helping to develop local planning and research capabilities in developing countries

b) Providing funding for feasibility studies and trial projects

c) Establishing training programmes

d) Setting up information networks.

Once biomass systems have been proven and large scale implementation schemes are begun, the requirements for capital will increase substantially. Most developing countries are likely to be reliant on outside sources of finance to a high degree. The industrialised countries and the aid agencies concerned must be responsive to these needs, and flexible enough to allow developing countries to find their own

II International Level (contd)

solutions to their respective energy problems.

11. Although the majority of applied R and D into biomass energy systems can only be performed effectively within the developing countries, the industrialised nations can help considerably in the area of basic research. Research which is directed solely at solving the energy problems of the developed countries, however, will tend to be of limited value to the developing countries, owing to the very different sets of conditions for which it will be designed. Indeed, there is a danger that as in virtually every other area of technological development the imbalance in research activities between developed and developing countries will lead to a situation where capital-intensive, high technology biomass options are the only ones which are readily available on the world market. Directing research specifically to the problems of the developing countries is, therefore, vital if this situation is to be avoided.

12. At the international level, there are a number of important opportunities for constructive collaboration between countries in developing biomass systems. These should be encouraged wherever possible. They include collaboration in information sharing, research and development, manpower training and technology transfer. Cooperation between developing countries with similar problems could be particularly fruitful in this context, particularly if the industrialised nations are slow to respond to the calls for increased technological and financial assistance from the poorer nations.

13. There is also considerable scope for improved collaboration between the international aid agencies involved in energy work. This would help to reduce duplication of effort and would increase the efficiency of utilisation of resources.

Format

This book is divided into seven chapters, each of which is as far as possible 'self-contained'. Whilst this format creates a certain amount of repetition, this is thought to be preferable to frequent cross-referencing in a review of this nature. A representative list of literature references is presented at the end of the main text.

The contents of the chapters are summarised as follows :

Chapter 1 Introduction

An introduction is given regarding the energy problems of the developing countries, and the background and objectives of the book are presented.

Chapter 2 The Present Role of Biomass Energy

This chapter outlines the present role of biomass energy in the developing countries. The characteristics of biomass utilisation are discussed in its major applications for household energy, agriculture and industry. A summary is given of the quantitative information that is available on biomass energy usage in developing countries.

Chapter 3 Energy Needs in Developing Countries

The causes and effects of energy problems in the developing countries are discussed. Energy needs are identified and categorised into three main classes : domestic energy needs, energy for development of rural areas, and energy for petroleum replacement.

Chapter 4 Biomass Energy Options

The main options for production and utilisation of biomass for energy are described. Production options are discussed in terms of presently available biomass resources, and the alternatives that exist for growing energy crops. Utilisation options are divided into conversion approaches (way of converting biomass to more convenient fuels) and end-use systems.

Chapter 5 Overall System Options

This chapter highlights some of the overall biomass system options (production, conversion, end-use) that appear to have the greatest immediate potential and widest applicability in developing countries - as judged on technical and economic grounds.

Chapter 6 Implementation

The experiences of biomass energy implementation efforts to date are summarised and some of the underlying social, political and practical problems involved in implementing biomass schemes are discussed. Comments are made concerning the need for infrastructures and extension servies and a summary is given of the factors that appear to be necessary requirements for effective implementation.

Chapter 7 Overall Considerations

This final chapter includes some general observations regarding the relationship between biomass energy development and overall development policy. The main areas where further research is needed are pointed out and opportunities for constructive collaboration are discussed. In conclusion, a number of comments are made concerning the role and potential for biomass energy in the developing countries.

Appendix A Data on Biomass Energy Use

A detailed summary is presented of the available data on biomass energy use in the developing countries, from village and national-level surveys.

Appendix B Institutions Involved in Alternative Energy Research

A partial listing is given of institutions and organisations that are actively involved in alternative energy research.

Alphabetical Index of Main Topics*

* Listings indicate the main areas in the text where these topics are discussed.

List of Figures

List of Tables

Chapter 1

Introduction

1.1 Background

The rising price of oil over the last decade has created very obvious difficulties for all countries reliant on oil for a major proportion of their commercial energy requirements. These problems seem certain to increase in the future as diminishing reserves and growing political pressures combine to force up the price and reduce the availability of oil. The pressures placed by oil price rises on the developing countries has been particularly severe since, although their consumption is only modest compared to the industrialised nations, they are often in the weakest position with regard to their ability to absorb energy price rises.

As a result of these problems, and because increased energy consumption is an unavoidable prerequisite of future economic development, the need to develop alternative energy sources (renewable and non-renewable) is clear. Among the various energy alternatives the development of energy from biomass is one area which has considerable promise in the developing countries and is receiving increasing attention.

Biomass energy is a general term that refers to the energy that can be derived from plant and animal materials, through a variety of conversion and end-use processes. Although many new approaches for using biomass for energy have been developed in the last few decades, by far the majority of biomass energy-use today occurs in the consumption of so-called 'traditional fuels' in the developing countries. These traditional fuels (wood, charcoal, crop residues and animal dung) are used primarily for cooking and heating in the home, but also play an important role in certain industries. It is in the safeguarding of these traditional fuel supplies that the second major challenge for biomass energy development arises.

The problems of traditional fuel supplies have only recently come to widespread public attention but they pose threats equally fundamental to the future prospects of many developing countries as does the well-known oil crisis. Like many of the problems of the developing countries, this 'second' energy crisis is closely related to population growth and the pressures that this puts on the environment for provision of both food and fuel. The widespread and rampant deforestation that afflicts many developing countries, and its consequences (increased flooding, erosion and siltation), are the most dramatic symptoms of this problem. It is also reflected, however, in many less obvious ways, such as the removal of tree cover from within agricultural areas (which is rarely recorded in forestry statistics) and the gradual shift to more and more ecologically damaging fuel collection practices (such as the increased use of crop residues and dung - which diverts valuable nutrients and organic matter away from the soil). In human terms the efforts required to provide fuel for basic needs are already a major burden for a large part of the world's population. In some countries these problems have reached crisis proportions - with people (commonly women and children) being forced to trek many miles to gather fuel, or spend large fractions of the total incomes on buying fuel. To alleviate these problems, and to ensure basic fuel (and food) supplies to future generations, it is essential that new methods are developed which can provide adequate fuel supplies in an ecologically sound and sustainable fashion.

A cursory examination of the fuel gathering and fuel-use practices presently employed in the developing countries shows them to be highly inefficient and wasteful (although they also have the overriding advantage of requiring very little financial expenditure) and points towards the significant improvements in the fuel supply situation that could be made by more efficient use of present biomass resources. Biomass supplies could also be increased in many cases, using relatively simple ploys such as planting trees on underutilised land. Using these, as well as other approaches, there appears to be considerable scope for using biomass energy systems to solve the domestic energy problems of the developing countries.

1.2 Objectives

The recognition of the potential that biomass energy has for helping to solve the energy problems of the developed countries has led to growing research activity in this area and an increasing involvement from governments as well as national and international organisations. Numerous studies have been made in the past few years of energy needs, available technologies and problems of diffusion of these technologies. It is a very broad subject, however, and both its technical diversity and the linkages it has with a wide variety of non-technical considerations (environmental, sociological, political and economic) make evaluation of the role of biomass energy development in overall planning an exceedingly complex and involved task.

The objective of this report is to draw together the many aspects of biomass energy development as they relate to the problems of the developing countries. This includes the technologies themselves (which range from very simple devices such as improved cooking stoves, to highly sophisticated options for the production of liquid fuels from various forms of biomass), the criteria for selecting technologies, the problems involved in implementing biomass schemes (and possible ways around these), the social and economic effects of introducing new energy use methods, and the role of biomass energy development in overall development planning.

The main thrust of this report is not to simplify the problems involved in biomass energy development, or to make judgements as to the best technologies or implementation approaches to use. Nor is any comprehensive attempt made to quantify the theoretical or practical feasibility of particular biomass options. Rather, the emphasis is placed on pointing out the issues that must be considered in arriving at an effective policy on biomass development and on highlighting the interactions, variables and unknowns that impinge upon this assessment process. A recurrent theme, therefore, is the differences that exist between developing countries, as well as within countries, and the important influence these have on the choice and viability of particular biomass systems.

Overall, no attempt is made to arrive at any sweeping conclusions as to the potential for biomass energy, either in specific cases or for the developing countries in general. This must be worked out in a case-by-case fashion, and with careful reference to a whole range of factors - many of which can only be assessed through practical experience. It is hoped, however, that the broad-ranging analysis presented here will provide a useful background to policy makers involved in the evaluation and development of biomass energy systems.

Chapter 2

The Present Role of Biomass Energy

In assessing the future prospects for biomass energy in the developing countries it is important to understand the major role that biomass energy already has in these countries. As a background to the discussion of biomass development options that is presented subsequently, this chapter provides an outline of the present role of biomass energy in the developing countries. An approximation of the global magnitude of biomass energy use is presented initially. This is followed by a general description, in qualitative terms, of the characteristics of biomass utilisation - what types of biomass are used for fuel in the developing countries; how these fuels are collected; how they are used and the functions that they perform. Finally, a summary is presented of the available quantitative data on biomass energy usage.

2.1 Global Perspective

The role that biomass plays in the current world energy supply has only recently been fully recognised. Although estimates are still only crude - due to the non-commercial nature of most biomass fuels and the resulting problems in measuring their consumption - energy from biomass probably accounts for as much as one seventh of total world energy consumption. (This is approximately four times the amount which is provided by hydroelectricity and nuclear power combined.) This startlingly high figure arises from the fact that the main biomass fuel forms - wood and charcoal - are the primary cooking and heating fuels for more than half of the world's population.

The role of biomass in the developing countries is particularly striking. In many countries, as is described later, biomass provides more than half of total energy needs. In extreme cases this reliance can be as great as 95%. The global distribution of biomass energy use between the developed and developing countries is shown diagrammatically in Figure 2.1. Whereas biomass represents only a minor component in the energy budget of developed countries, on average it provides for approximately 43% of the energy use of developing countries. The majority of this use occurs in rural areas, where about 70% of the population lives, and where biomass fuels are often the only energy source which is both available and affordable to the mass of the population.

Figure 2.1 Global Distribution of Energy Use (1978)

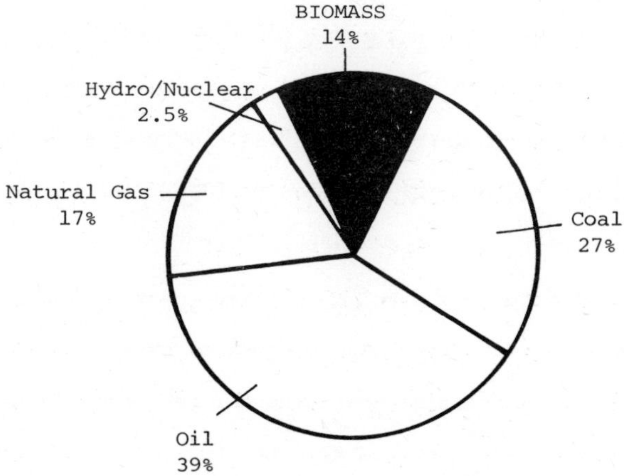

WORLD

(TOTAL = 300 x 10^9 GJ)

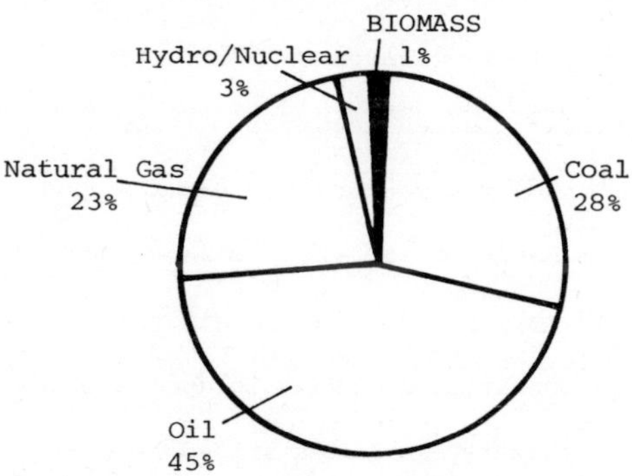

DEVELOPED COUNTRIES

(TOTAL = 208 x 10^9 GJ)

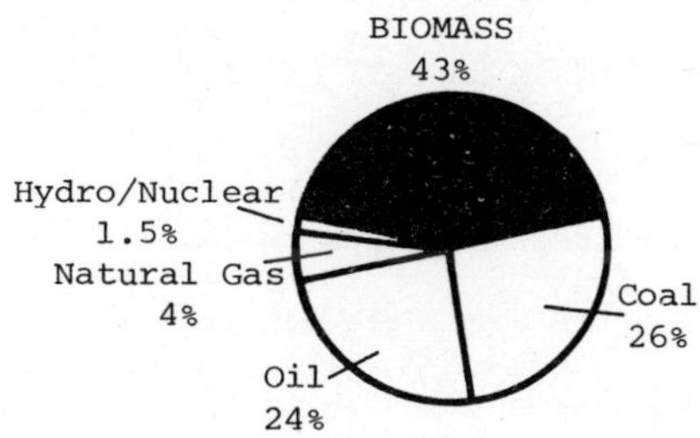

DEVELOPING COUNTRIES

(TOTAL = 92 x 10^9 GJ)

Source : Commercial Energy - UN; "Energy Supplies 1970-78"
Biomass Energy - Calculated assuming annual per capita consumption of biomass fuels in developing countries equals 15 GJ for rural populations (2.1 billion people) and 8 GJ for urban dwellers (0.9 billion people). Figures for developed countries taken from Parikh (reference 58).

2.2 Characteristics of Present Biomass Energy Utilisation

The main forms of biomass used for energy purposes in the developing countries are the so-called 'traditional fuels' - wood, charcoal, crop residues and anumal dung. These are mainkind's oldest energy sources and their collection and use remains an integral part of the fabric of life within the developing countries, particularly in the rural sector.

The pattern and extent of biomass energy utilisation varies significantly between developing countries, reflecting differences in energy needs, fuel-use practices and the relative availability and cost of biomass fuels compared to other fuel forms. Within countries, energy use patterns are also far from uniform. They commonly show regional variations as well as difference between income groups. The types of biomass used can also vary at different times of the year, due to seasonal fluctuations in biomass availability.

The overall pattern of biomass energy use in developing countries is extremely complex, being closely linked to agricultural practices as well as various cultural, social and economic factors. Only in a very few cases have the micro-level features of biomass use been characterised in detail. On the basis of the sketchy knowledge that is available at present, however, the main characteristics of biomass energy use in the developing countries in its main applications - for household uses, agriculture and industry - are summarised below.

a) Household Energy

Use of biomass for cooking and heating in the home is the largest application of biomass energy. Wood is the commonest biomass fuel used, followed by charcoal but dung, crop residues and vegetable wastes are also important in many countries where wood is not readily available. These fuels are burnt on open fires or in simple stoves of various designs, usually with very low efficiency (the nature of cooking stoves is discussed in more detail in subsequent sections). The use of commercial fuels (kerosene, bottled gas, electricity) for cooking and room heating in the developing countries, although it has been increasing over the years, is still relatively rare and is usually limited to the more affluent sectors of the population, largely in urban areas. (It is interesting to note that traditional fuels, especially charcoal, are sometimes preferred to the more modern fuel forms for certain types of cooking, owing to the flavour they impart to the food.)

In rural areas biomass fuels are generally collected on a 'free good' basis, usually by women and children within the community. The types of biomass used and the effort required to gather them depends on availability. Where population pressures are not too severe, fuelwood can usually still be found relatively easily - from wooded land, scrub and isolated trees that exist near to the home. If wood is scarce, on the other hand, people have to be less discriminating the the types of wood they use and they are forced to travel further afield to collect it; increasing the 'social cost' of the fuel significantly. In some regions of the world this can mean journeys of half a day or more, representing a very major burden on the rural community.[56]

Conversion of wood to charcoal tends to be more common when transportation distances are long. The conversion methods used are generally fairly basic, employing either covered wood-piles or simple earthen kilns, both of which are very inefficient. Charcoal making also occurs in some cases where wood is being used locally, such as in Thailand and Senegal, due to a preference for charcoal over other cooking fuels. In most cases this implies a greater overall consumption because although charcoal stoves may be two to ee times as efficient as wood stoves, the initial charcoal making process commonly involves energy losses of as much as 80% of the original energy content of the wood.

In cases where adequate wood supplies cannot easily be found and commercial fuels are either too expensive or are unobtainable, crop residues are often employed for fuel. These include stalks and husks from various grain crops, cotton and nut wastes, and sugar cane bagasse. In some countries, notably Bangladesh, these residues are the major source of fuel for domestic uses.[16] More commonly, however, their use is restricted to the times of the year when they are readily available (following harvest, for example).

In many developing countries the availability of crop residues for energy purposes is severely limited by the fact that these residues are also used extensively for animal feed, as well as for various other purposes (such as thatching). Under such circumstances, often the only other locally available fuels are animal dung - and to a lesser extent, vegetable wastes. Dried dung cakes are an important source of fuel in parts of India and Bangladesh, for example, as well as in the sparsely forested Andean regions of Bolivia and Peru.[89] The burning of animal dung has also been reported in some Sahelian countries, parts of the Transkei in South Africa,[8] and in Ethiopia and Iraq.[50] On a world-wide basis, it has been estimated that between 500 million and one billion people rely on animal dung for a portion of their domestic energy needs.

In urban areas charcoal is the predominant household fuel, with wood playing a less important role. To satisfy urban fuel needs, wood and charcoal supply industries exist in most countries. In some cases these are fairly well organised. For example, in Thailand, mangrove forests are managed in a semi-controlled fashion for charcoal production. In others, fuelwood is extracted as part of organised lumbering operations and is converted to charcoal in a reasonably efficient fashion. More commonly, however, wood is collected on an opportunist basis from wherever it can be obtained, with very little regard to long-term forest productivity. Often this is performed by individuals, frequently operating illegally, who convert the wood to charcoal using inefficient small-scale kilns and sell it to traders who transport the wood to the towns. Frequently these 'middle men' make substantial profits from this trade since the urban demand is very inelastic (particularly in the context of the rising cost of oil-based alternatives) and because the charcoal producers - who are typically the poorest sectors of the rural community - have very little bargaining power.

As a consequence, charcoal prices in urban areas can be quite high, and often represent a major fraction of the income of the poorer sectors of the population. Alternatively, individuals must scavenge for whatever fuels they can find locally, or travel out into the surrounding areas to collect fuels themselves. Such a daily exodus is a common feature of urban life in a number of developing countries, as is the resultant denudation of trees and scrub for large distances around urban areas.

b) Agriculture

Biomass is an important source of energy for agriculture in many developing countries in that it provides the food eaten by animals and humans. Often draught power and human effort are the only mechanical inputs into agriculture and as such, are a vital part of the farming system. Food for draught animals is usually provided from a variety of sources, depending on availability and competing uses. These include crop residues at certain times of the year and leaves, bark, herbage and other miscellaneous vegetation at other times.

c) Industry

Wood and charcoal are commonly used for industrial purposes in developing countries. Most of these operations are relatively small-scale, but in a few cases, such as the Brazilian steel industry (which is predominantly fuelled by charcoal), large-scale biomass-using facilities exist. The major industries employing wood and charcoal are as follows :

- tobacco curing
- brick-making
- rubber manufacture
- brewing
- pottery making
- sugar production
- metal extraction
- steel making
- blacksmiths
- food drying and curing

To satisfy the fuel needs for these activities, various types of commercial and semi-commercial wood and charcoal operations exist. Often these are an extension of the informal and uncontrolled fuel supply industries described above, that provide wood and charcoal for domestic needs. In a few cases, however, managed fuelwood plantations exist - such as in Brazil where approximately 8% of the 2.3 million tonnes of charcoal used annually by the iron and steel industry (1971) comes from eucalyptus plantations.[80]

In a number of industries, wastes produced as part of the process are used as fuel. The extensive use of a sugar-cane bagasse as a fuel for sugar production is the largest example in this category. Others include the use of sawdust, bark and wood shavings by sawmills, and the burning of various organic wastes by the food processing industry.

2.3 Summary of Energy Use Surveys

Understanding the patterns of energy use at both local and national levels are an essential part of formulating effective energy policies. Reasonably good data exists relating to the consumption of commercial fuels in developing countries but the sparseness and unreliability of information on biomass utilisation is a major planning impediment. This problem arises primarily because of the non-commercial nature of most biomass fuels. They are often collected by the user and rarely pass through easily monitorable channels. For this reason, most official statistics on biomass energy use have been, and still are, gross underestimates of actual consumption levels.

To provide accurate information on the use of biomass fuels detailed field surveys are required. These are very laborious in practice, owing to the many inherent complications. This is particularly true for household energy studies, which are hampered by the following problems :

- the seasonality of fuel supplies and the resulting variation in consumption during the year
- the wide variety of fuels used and their different moisture and heat contents
- estimating the efficiency of conversion of wood to charcoal
- the variable efficiency of stoves
- the problems of relying on personal recollection for assessing how much fuels are used
- the tendency for people interviewed to tell survey teams what they want to hear rather than the true facts
- variations in energy use within an area due to occupational and income differences as well as other factors

A number of attempts to measure biomass energy consumption have been made, however, and a summary of the results from these surveys is presented in Appendix A. In Table 2.1 a selection of data is shown from some of the more detailed village level energy surveys that have been performed. As can be seen, the quantity of biomass used varies significantly between countries. In the Bangladesh study, for example, the biomass energy consumption (6.7 GJ/capita/year) is only about one quarter of that reported for a village in the Peru (27 GJ/capita/year). Whilst measurement errors and methodological differences may contribute somewhat to this wide variation, they certainly cannot be the only cause. Differences in climate, fuel availability and cooking methods are presumably the main reason behind this variation, as well as factors such as the availability and price of commercial alternatives.

Overall, the amount of biomass-derived energy consumed annually by rural inhabitants in the developing countries probably averages about 15 GJ per capita. This is equivalent to 1 tonne or 1.4 m^3 of air-dry wood.

The energy consumption of urban inhabitants in the developing countries has received less attention than the consumption in rural areas. Since charcoal is the predominant fuel for most urban dwellers, problems arise in measuring how much primary energy is used (in this case wood) due to uncertainties regarding the efficiency of the charcoal making process. From the limited number of surveys that have been performed, however, it appears that biomass energy use is generally somewhat lower than in rural areas[19] - perhaps 8 GJ/capita/year on average (equivalent to 0.5 tonnes or 0.8 m^3 of wood). One can presume that this reduced consumption is due partially to the generally higher

TABLE 2.1 Summary of Selected Village Energy Surveys

<table>
<tr><th>COUNTRY (Source)</th><th>VILLAGE or REGION</th><th>WOOD</th><th>DUNG</th><th>CROP RESIDUES</th><th>OTHER</th><th>TOTAL BIOMASS</th></tr>
<tr><td>Bangladesh (Briscoe, 1979)</td><td>Ulipur</td><td>1.4</td><td>0.2</td><td>3.8</td><td>1.3</td><td>6.7</td></tr>
<tr><td>India (Ravindranath, 1980)</td><td>Ungra</td><td>8.1</td><td colspan="2">0.7</td><td></td><td>8.8</td></tr>
<tr><td>Nepal (Hughart, 1979)</td><td>Hill Region</td><td>11</td><td></td><td></td><td></td><td>11</td></tr>
<tr><td>South Africa (Best, 1979)</td><td>Transkei</td><td colspan="2">5.3</td><td></td><td></td><td>5.3</td></tr>
<tr><td></td><td>Kwazulu</td><td>17</td><td></td><td></td><td></td><td>17</td></tr>
<tr><td>Upper Volta (Ernst, 1978)</td><td>Ranga</td><td>9</td><td></td><td></td><td></td><td>9</td></tr>
<tr><td></td><td>Boulenga</td><td>12</td><td></td><td></td><td></td><td>12</td></tr>
<tr><td>Kenya (Mun'gala, 1979)</td><td>Machakos</td><td>15</td><td></td><td></td><td></td><td>15</td></tr>
<tr><td>Iran (Vojani, 1978)</td><td>Semnan</td><td>18</td><td></td><td></td><td></td><td>18</td></tr>
<tr><td>Peru (Winterhalder, 1974)</td><td>Quechua</td><td></td><td>27</td><td></td><td></td><td>27</td></tr>
<tr><td>Mexico (Makhijani & Poole, 1975)</td><td>Arango</td><td>13</td><td></td><td></td><td></td><td>13</td></tr>
</table>

All figures in GJ/capita/year

Source : See Appendix A for details.

income in urban areas, and the greater access to commercial fuels. For poorer members of the community, though, conservation must also be an important factor since - to them - fuel is an expensive commodity.

Estimates of energy inputs to agriculture, through human and animal power, are very crude. This is due to the problems involved in defining and measuring the amount of food energy converted to useful work. The results of several attempts to do this are summarised in Table 2.2. These underline how animate energy of this form can be a major component in the overall energy picture, typically accounting for as much as 10% of total consumption in rural areas and often representing the only major energy input into agriculture.

Results of national energy studies are shown in Table 2.3. Again, only selected examples are shown but these indicate the range of energy use structures that exist. Thus, while some of the more advanced developing countries now rely heavily on oil and other commercial fuels, the less developed countries are almost totally dependent on biomass for energy. A more detailed list of national energy use statistics is given in Appendix A.

A general trend that emerges from these figures is that although total energy use varies significantly between countries this is mainly due to different levels of consumption of commercial fuels. Per capita use of biomass-derived fuels, on the other hand, is fairly constant. This underlines the importance of biomass fuels for providing basic energy needs and is indicative of their continuing role as the major source of energy for domestic uses (particularly in rural areas), even in countries where modernisation and industrialisation have led to a large growth in fossil fuel consumption.

TABLE 2.2 Energy Inputs in Rural Areas from Human and Animal Work

Country	Village	Human Work	Animal Work	Total Energy Consumption	Percentage of Total Energy supplied by human and animal work
		- GJ/capita/year -			
China[2]	Peipan	1.0	1.4	26.7	9%
Tanzania[2]	Kolombero	0.9	-	23.8	4%
Nigeria[2]	Batagawara	0.9	0.2	16.8	7%
Mexico[2]	Arango	1.1	2.0	56.0	6%
Bolivia[2]	Quebrada	1.1	2.8	38.6	10%
Bangladesh[1]	-	1.0	1.5	7.9	32%
India[1]	-	1.0	1.5	10.8	23%
India[3]	Pura	0.7	0.1	10.8	7%
India[4]	Ungra	0.9	0.4	10.6	12%

Sources : 1. Revelle (1976) Science 192, p.969, Table 1.

2. Adapted from A. Makhijani and A. Poole (1975) "Energy and Agriculture in the Third World" by Revelle using the assumptions described below.

3. A.K.N. Reddy and D.K. Subramanian (1979) "The Design of rural energy centres" Proc. Ind. Acad. Sci., Vol.2, part 3.

4. N.H. Ravindranath et al (1980) "An Indian village agricultural ecosystem - a case study of Ungra village, Proc. Conf. "Energy from Biomass", Brighton, UK.

Note : Figures for human and animal work are an approximation calculated using the method of Revelle in which only the energy expended in performing work is counted. This is assumed to be 33% and 40% of the energy content of the food eaten by humans and draught animals, respectively. If the entire food energy content was included, figures would be substantially higher.

TABLE 2.3 Summary of Selected National Energy Consumption Surveys

Country	Commercial[1] Energy	Biomass[2] Energy	Total Energy	Percentage of Energy from Biomass
	- GJ/Capita/year -			
Bangladesh	1.2	3.0	4.2	71%
Niger	1.1	8.0	9.1	88%
Gambia	3.1	7.0	10.1	69%
Morocco	8.4	2.3	10.7	21%
India	5.2	6.0	11.2	54%
Ethiopia	0.6	11.7	12.3	95%
Nepal	0.3	13.5	13.8	98%
Somalia	2.9	15.0	17.9	84%
Bolivia	10.7	8.3	19.0	44%
Sudan	5.0	20.0	25.0	80%
Thailand	9.6	16.5	26.1	63%
Tanzania	1.9	25.5	27.4	93%
China	24.5	10.0	34.5	29%
Brazil	23.2	11.7	34.9	34%
Mexico	40.5	4.0	44.5	9%
Libya	55.3	3.0	58.6	5%
Developing Countries (Average)	17.3	13.1	30.1	43%

Sources : 1. UN (1979) World Energy Supplies 1970-1978. Figures refer to 1978 consumption

2. Selected data taken from Biomass Energy Surveys presented in Appendix A. Figures refer mainly to wood and charcoal consumption; see Appendix A for further details.

Chapter 3

Energy Needs in Developing Countries

The present energy problems of the developing countries can be separated into two main categories; the well known 'oil crisis' that affiicts all oil importing nations and the much less well appreciated but equally important 'second energy crisis' resulting from the diminishing supply of traditional fuels - wood, dung, and crop residues. The nature of these problems is discussed in this chapter and some conclusions are drawn as to what are the major energy needs of the developing countries.

Energy use patterns within developing countries form a complex and interrelated system, in which variations in the supply of one fuel affects the demand for others. Whilst the system must be viewed as a whole in overall energy planning, the energy needs of most developing countries can be divided into three basic categories, as follows :

- Domestic energy needs
- Energy for rural development
- Energy for petroleum replacement

The reasons why these needs have arisen and general observations as to their size and nature, are discussed below.

3.1 Domestic Energy Needs

As described in the previous section, the vast majority of rural populations in developing countries, as well as a significant number of city dwellers, have traditionally relied on biomass fuels for cooking and heating within the home. The commonest fuels used are wood and charcoal, although animal dung and crop residues are important in some countries. The predominant utilisation method is direct burning, either on open fires or in simple stoves, both of which are generally very inefficient.

Whilst these traditional methods have been used successfully for thousands of years, in many regions of the world today the system is breaking down and supplies of fuel can no longer meet demand in an ecologically sound and sustainable way. Because of the imperative for providing domestic energy, however, fuel must be found regardless of consequences. Collection methods are being adopted which, whilst they satisfy the short-term need for fuel, inflict permanent damage to the environment, thereby reducing its ability to provide further fuel in the future.

The severity of domestic energy problems varies significantly between countries. Some have sufficiently sparse populations and abundant resources as to be relatively immune from these problems, at least in the near term. In others, fuel shortages have become, or are soon to become, so acute that they are a dominating consideration in daily life. The chief symptoms of fuel supply problems are as follows :

a) Depletion of woodlands in and around population centres and an increase in the distance that must be travelled, as well as the time and effort required to find fuelwood

b) Increased encroachment into forest areas for fuel collection and charcoal manufacture and the resulting effects that this has on speeding up deforestation rates.

c) Increase in price of wood and charcoal, particularly in urban areas

d) Increased utilisation of agricultural residues and animal dung for fuel - which has the effect of diverting nutrients and organic matter that could otherwise be returned to the soil.

The underlying causes of these problems is not always easy to pinpoint, however. While they are obviously related to population growth and the pressure that this places on the environment for the provision of fuel, there are often a number of other important factors involved.

Deforestation, for example, whilst inevitably aggravated by the excessive cutting of trees for fuelwood, can rarely be attributed solely to this cause.[22] The cutting of forests to open up new agricultural land is often an equally important factor. Overgrazing by animals and uncontrolled timber extraction can also be major contributors. In such cases, an energy policy which concentrates only on the firewood aspect of deforestation is unlikely to be effective in solving the real problem. In a similar way, if unequal ownership of existing fuel resources is the main reason behind the fuel supply problems of the poor, it must be recognised as such so that policies can be devised accordingly.

To alleviate current fuel shortages and to ensure fuel supplies to future generations, sustainable and environmentally sound energy alternatives are urgently required. For the majority of the poorer sectors in the developing countries, locally grown biomass fuels are the only source of energy which is within their financial reach. This situation is unlikely to change in the near future since virtually all non-biomass alternatives - kerosene, bottled gas, etc - are much too expensive to have any widespread impact unless they are heavily subsidised. Developing low-cost methods of producing and using biomass for energy is therefore of the highest priority.

3.2 Energy for Rural Development

Providing a sustainable and affordable energy source is an important component in rural development.* Coupled with other inputs - financial, technological, organisational, etc. - a relatively small increase in energy use can permit substantial improvements in agricultural productivity and can allow growth in rural industries and public services.

At various levels of modernisation,energy requirements range from such basic needs as draught power during periods of peak agricultural activity to, at the other end of the scale, electrical energy to run refrigerators and small industries. As oil prices rise the incentives for developing locally available energy alternatives to supply these needs is increasing.

---

* Energy is also important for development of urban areas - for industry, transportation systems, public services, etc. These needs are not dealt with separately in this report since they largely overlap with the needs for petroleum replacement. Some of the technologies described for meeting rural energy needs will also be applicable in urban areas, however, and the classification system used is not intended to preclude this option.

3.3 Energy for Petroleum Replacement

The rising price of oil and the growing demand, combine to place severe strains on the economies of all developing countries reliant on imported oil for a major proportion of their commercial energy requirements. Whereas in the early 1970's oil import bills typically accounted for 10% or less of total export earnings, in many developing countries today 40-50% of foreign exchange earnings must be spent to finance oil imports. As a result, many countries are now facing severe balance of payments problems and those able to borrow money to pay for oil imports are incurring unmanageable debts. In view of these problems, the need to develop indigenous energy alternatives to replace imported oil is both obvious and urgent.

Looking at the options for replacing petroleum, it is relevant to consider the pattern of present oil consumption and the historic trends in energy use. Originally, developing countries relied mainly on wood and other traditional fuels for their modest energy needs. However, as development and westernisation brought industry, improved transportation and other innovations to these countries, cheap foreign oil began to have a more and more important role. This increase in oil consumption has been particularly dramatic in the last 30 years, as shown in Figure 3.1, and has risen by a factor of 10 during this period in absolute terms and by a factor of 5 on a per capita basis.

Major differences exist between countries. In countries where economic growth has been slow, present oil consumption levels are still very low. This does not imply that they are immune from the oil crisis. On the contrary, it is often the poorest nations that suffer most, because of their inability to pay higher oil prices and their low credit rating on the international loan market. The range

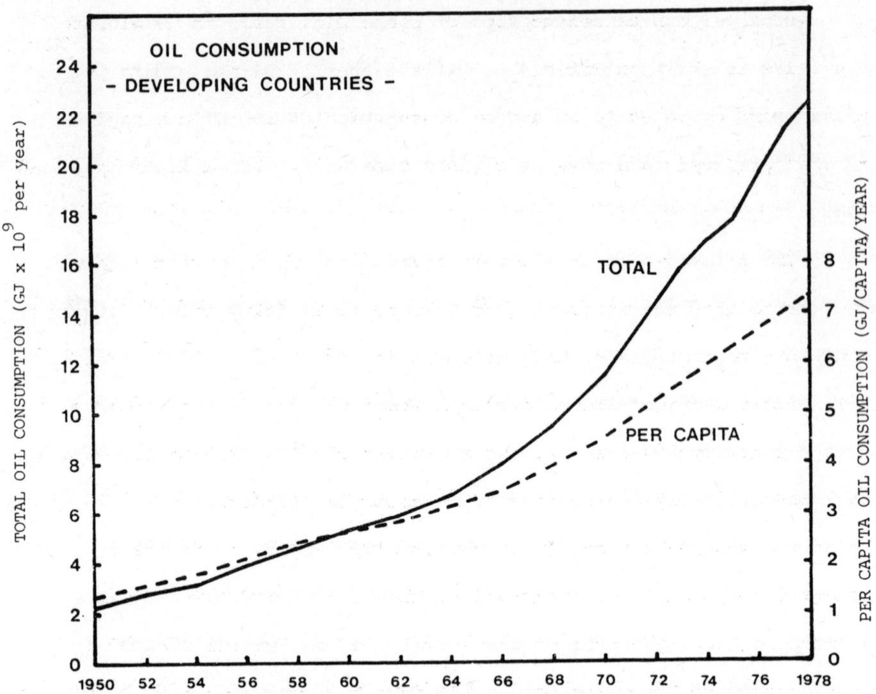

Figure 3.1 Trends in Oil Consumption in the Developing Countries since 1950

Source : UN (1979) World Energy Supplies 1970-1978

of present per capita consumption of commercial fuels in developing countries is shown in Table 3.1. It averages 17 GJ per capita per year, which compares to an annual consumption of 190 GJ per capita in the developed countries as a whole and 333 GJ per capita in the USA.

The breakdown of this total figure into the main commercial fuel forms (for the developing countries) is as follows : oil 43%, coal 46%, natural gas 8% and nuclear/hydro 3%.

These average figures overemphasise the importance of coal for most countries, however. Of the approximately 800 million tonnes of coal used annually in the developing countries (1978 figures - reference 88), more than 90% is consumed by only four countries; namely China, India, N. Korea and S. Korea. China, alone, consumes as much as three quarters of this total. If one subtracts the contribution of these four countries from the aggregate figures for the developing countries, the significance of oil as the main commercial energy source becomes much clearer. Thus, for the remainder of the developing countries, the breakdown is as follows : oil 72%, coal 7%, natural gas 17% and nuclear/hydro 4%. It should also be noted that, of the coal that is used, approximately 98% of it is produced indigenously in the countries in which it is used. It is, therefore, relatively immune from the forces that are driving up oil prices.

An additional feature of the present pattern of oil use in the developing countries that is relevant to the planning of energy alternatives is that of centralisation of demand. The trend of urbanisation and the tendency to locate industries in and around the major cities, has led to the situation where in many countries the majority of oil consumption is concentrated in just a few areas.

TABLE 3.1 Consumption of Commercial Energy in Selected Developing Countries (1978)

Country	Coal	Oil	Natural Gas	Nuclear & Hydro	Per capita consumption GJ/year
	- percentage contribution -				
Nepal	6%	81%	-	13%	0.3
Ethiopia	-	92	-	8	0.6
Mali	-	97	-	3	0.9
Bangladesh	10	51	36	2	1.2
Tanzania	-	94	-	6	1.9
Somalia	-	100	-	-	2.9
India	65	28	2	5	5.2
Guatemala	-	97	-	3	7.6
Indonesia	1	84	14	1	8.1
Bolivia	-	87	8	5	10.7
Zambia	29	42	-	28	13.9
Peru	2	85	6	7	19.0
Brazil	9	75	2	14	23.3
China	80	18	1	1	24.5
S. Korea	38	61	-	1	39.8
Mexico	8	68	20	3	40.6
Argentina	3	69	25	2	54.9
Libya	-	100	-	-	55.3
N. Korea	91	4	-	5	79.2
Venezuela	1	49	46	4	87.6
Developing Countries (Average)	46%	43%	8%	3%	16.8

Source : UN (1979) World Energy Supplies 1970-78.

This has major implications in the analysis of petroleum replacement options because the problems of transporting fuels - from where they are produced to where they are needed - can be considerable. This is particularly true for decentralised energy systems, notably those (such as biomass energy) which are solar based, and provides a strong incentive for locating new energy requiring activities as near as possible to the sources of energy they will use.

Overall, the replacement of oil in its current roles is likely to be a difficult task for almost all countries because present energy consumption patterns have grown up with cheap oil very much as their basis. Meeting this challenge, however, is an essential part of ensuring economic survival and permitting continued development of third world nations.

Chapter 4

Future Biomass Energy Options

4.1 Background

As described, biomass energy already plays a crucial role in the developing countries although population and ecological pressures are now placing this role in increasing jeopardy. A wide variety of options exist, however, for expanding the use of biomass energy and potentially these could have a major impact on all three categories of energy needs outlined in the previous chapter; domestic needs, rural development and petroleum replacement.

Theoretically, the choice of biomass energy systems is huge. Possible biomass sources vary from scavenged crop residues to large, intensively managed energy plantations. Likewise, utilisation approaches range from direct burning in simple stoves to thermochemical conversion to synthetic petroleum. In judging the applicability of these various alternative it is vital that they are assessed in the context of the local conditions and constraints under which they will be used. These differ widely, both between and within countries, and mean that while technically advanced and high efficient technologies may be appropriate in some circumstances, simplicity and low cost are the dominant considerations in others.

Thus, in identifying appropriate energy paths, it is vital that each of the following questions is fully considered :

1. The magnitude and nature of energy needs, present as well as future, and the possible differences that can exist between what is theoretically needed and the demands that exist in practice.

2. The existing biomass resource base and the resource development potential, particularly in view of the complex agricultural and ecological constraints on the system.
3. The technical feasibility of biomass utilisation options in view of the size and location of available biomass resources and the readiness of technologies for implementation.
4. The economic resources available for investment in energy production systems and for purchasing the derived fuels.
5. The levels of available skills and the existence of infrastructures necessary for the introduction and operation of new energy systems.
6. The sociological environment, the problems of acceptance of new technologies and the resulting effects on the social system.
7. The legal and governmental barriers to the implementation of new energy policies.
8. The environmental pros and cons of developing biomass resources.

In this chapter the main biomass energy options are introduced, both for production of biomass and for its utilisation. For the sake of clarity, the emphasis if largely technical and an attempt is made to describe the basic principles behind the various biomass technologies - and what their theoretical advantages and disadvantages are. In later chapters these options are discussed with more specific reference to the conditions within developing countries and the various contraints and limitations on their use.

4.2 Biomass Production

Globally, the amount of biomass produced each year through photosynthesis is vast. It has been estimated that net annual photosynthetic production of organic matter amounts to roughly 2×10^{11} tonnes; equivalent to 3×10^{21} J in energy terms.[36] This is approximately ten times present world energy consumption.

This apparent abundance of plant energy is somewhat misleading, however, because of the many limitations that exist relating to its use. These include geographical factors, economics, questions of ownership, and competition for other uses - especially for food. Renewability is also a key consideration. If harvesting a particular biomass source results in degradation of the ecosystem that produced it then its availability will be only temporary and serious and permanent environmental damage may result if it continues. Indeed, it is this very practice of meeting short-term energy needs in ways that jeopardise long-term ecosystem productivity that poses such a threat to future supplies of biomass energy in the developing countries.

The first step, therefore, in assessing the potential for biomass energy is to determine the amount of biomass which is available in practical terms. This is discussed in Section 4.3.1. Following this, the main options for increasing biomass productivity through growing plants specifically for energy purposes, as 'energy crops', are outlined.

4.2.1 PRESENTLY AVAILABLE BIOMASS RESOURCES

Presently available biomass resources can be divided into five main categories; wood, crop residues, animal manure, household wastes and food processing wastes. The abundance of these resources varies tremendously between countries depending on factors such as local agricultural practices and crop productivity, and the amount of forest and woodland remaining. Quantitative data on this subject is generally very scarce, however, and most of the estimates used in this section are calculated from rather crude national statistics on forestry and agriculture.

The extent to which present resources are available in practice to the population, and their adequacy for meeting current needs, is even more difficult to assess without exhaustive local surveys. Obviously, the distribution and density of population are important considerations in this context as is the efficiency of the current energy use practices. Inequality of access to energy sources can also play a central role and taken to an extreme can mean that poorer members of the community can be starved of energy (as well as food) while their more affluent neighbours have food and fuel to spare.

Given, therefore, that biomass availability is very site-specific, a description of the main resource categories is presented below. The chief constraints on the use of existing resources are summarised in Table 4.1.

a) Wood

Although deforestation has taken a severe toll in many regions of the world, all developing countries still have some wood resources remaining and in most countries they represent the largest single renewable energy resource. These wood resources can be divided between those which exist within the major forest areas and the smaller category of dispersed woodlands. Accurate estimates of total national

TABLE 4.1 Limitations on the Availability of Biomass Materials for Use in Energy Purposes

Biomass Resource / Nature of Limitation	THEORETICAL LIMITATIONS	PRACTICAL LIMITATIONS	ENVIRONMENTAL LIMITATIONS	SOCIO-CULTURAL LIMITATIONS
WOOD	Amount of forest and woodland remaining and the levels of sustainable wood production	Distribution relative to population centres, collection problems, and competition for other uses.	The need for good collection practices to prevent forest damage.	Ownership of forests, poor forestry knowledge and inappropriate forest-use traditions.
CROP RESIDUES	Amount of residues produced - a function of crop types, yield, and area of farmland.	Competition for other uses and seasonality of supply.	Soil degradation due excessive removal of residues and inadequate nutrient recycling.	Inequalities in land ownership and rights to use crop residues.
ANIMAL MANURE	Amount of dung produced - a function of the number and type of animals, and levels of health.	Collection problems with free-ranging animals.	Removal of nutrients from the soil, unless measures are taken to recycle them.	Ownership of animals and cultural objections to collecting and using dung.
FOOD PROCESSING WASTES	Amounts of wastes generated - which depends on the quantities of food processed and the processing efficiency.	Competition for other uses and seasonality of supply.	May be an advantage in some cases where wastes present a disposal problem.	Ownership of wastes.

wood resources are generally unavailable and those that do exist nearly always refer to the major forest areas only. The many problems in defining and measuring forest areas and productivities are the primary cause of the lack of good wood resource data, and these are aggravated by the rapid rate at which conditions have been changing due to deforestation.

To provide an idea of the range of wood resources existing in developing countries, estimates of the forest area remaining in selected countries on a per capita basis are shown in Table 4.2. A rough approximation of the sustainable wood yield from these forests, expressed as GJ per capita per year, is also included (these figures are from a different source and take into account the variable productivity levels of different forest types). Such crude calculations indicate how in some countries annual production from existing forests represents a major biomass resource whilst in others it falls well short of meeting present wood energy requirements of from 5-20* GJ per capita annually - let alone the many non-fuel uses of wood.

It must be remembered, however, that even though very large amounts of wood may be produced annually in some countries, not all of this is actually available in practical terms for energy purposes. Disregarding ownership considerations for the moment, the fact that many forest areas are very remote tends to make the extraction of wood purely for energy purposes uneconomic.

The economics of cutting forests for energy can be substantially improved, however, if it is combined with logging in a joint operation. Residues from normal logging activities - which consist of branches, stumps, thinnings and diseased and unsuitable wood - can amount to a third or more of the quantity of lumber removed. Most of this is presently unused due to the inconvenience and transport costs of

* See Table 2.1 - Consumption in some countries may be higher

TABLE 4.2 Forest Resources in Selected Developing Countries

COUNTRY	POPULATION[1] (10^6)	FOREST AREA[1] REMAINING (10^6 hectares)	FOREST AREA[2] PER CAPITA (hectares)	SUSTAINABLE[3] WOOD YIELD (GJ/Capita/yr)
Congo	1	17	12.1	571
Brazil	116	350	3.0	350
Argentina	27	60	2.2	123
Burma	35	38	1.1	110
Tanzania	18	30	1.7	88
Thailand	48	20	0.4	37
Nepal	14	4	0.3	21
Mali	7	6	0.9	12
Upper Volta	7	4	0.5	10
India	694	75	0.1	7
Kenya	16	1	0.06	2
Bangladesh	89	3	0.03	2
Developing Countries (Average)[4]	3,000	2,222	0.74	30

Sources : 1. World Bank (1980) Developing Countries fuelwood supply/demand analysis
2. Calculated from previous columns
3. World Bank (1979), "Prospects for traditional and Non-Conventional Energy Sources in Developing Countries", Table I-4. Based on forest area data in 1977 FAO Production Yearbook and data on average forest yields in D.E. Earl (1975), Forest Energy and Economic Development
4. Calculated from 1979 FAO Production Yearbook and Earl (op.cit), Table 4.4.

Note : This data is almost certainly an overestimate of both forest area and productivity levels, but is the best available at present.

removing it. Conversion of these residues to charcoal using simple kilns (see Section 4.3.1) is a useful way of getting round these problems and is presently used in parts of East Africa, as well as a number of other countries. Furthermore, as Earl has noted,[25] careful removal of logging residues can actually have a beneficial effect on forest productivity, increasing the yield of both timber and fuel - although excessive use can obviously be environmentally detrimental.

A sub-category of forest resources with some energy potential are the residues produced by sawmills and pulpmills. The amount of residues produced from sawmills, in the form of sawdust, shaving and offcuts depends very much on the efficiency of the operation but often amounts to almost half of the timber processed. These provide a useful fuel for the powering of the mill and are frequently used for this purpose. Any that is left over can be employed for other uses. In the long run, however, improved technologies such as thinner blades will tend to reduce waste production as will increased utilisation of residues for the manufacture of composite materials.

Kraft black liquors, which contain the residual liquor and cellulose from the pulping process, are one of the major sources of fuel for the pulp and paper industries. At present they are incinerated although potentially they could be converted to more valuable fuels using more sophisticated conversion processes.

b) Crop Residues

In principle, crop residues represent a major energy resource but there are a number of fundamental limitations on their use. The amount of crop residues produced depends on a wide variety of local conditions but it is the relative importance of competing non-energy uses that determines their availability for energy purposes.

The primary factors influencing the amount of crop residues produced are the type and variety of crops planted and their subsequent yield. Crop yield, in turn, is a complex function of agro-climatic conditions and farming methods and can vary significantly from year to year. The range of crop residue productivities, on a per hectare basis, is shown in Table 4.3 for the main cereal crops grown in developing countries. The range is very large, both between crops and between countries. So, whilst a rice farmer who uses modern cultivation methods and multiple cropping could produce as much as 10 tonnes of straw annually, a farmer using traditional methods for growing a lower yielding crop (such as sorghum or millet) would produce less than one tonne of residues per hectare.

The significance of crop residues in energy terms is indicated in Table 4.4. Here, the per capita content of residues produced annually from cereal crops is shown for selected countries. These figures are derived from national agricultural production data based on a number of rough assumptions concerning crop-to-residue ratios and average energy content of residues. Whilst these figures are far from accurate (every component of the calculation being open to argument), they do provide an order of magnitude estimate of the size of the crop residue resource. Although there is a considerable variation between countries, on average, approximately 5.6 GJ of cereal residues are produced annually, per capita, in the developing countries. This could provide an important contribution to energy supplies in view of the fact that energy consumption levels of 5 to 20 GJ are typical in rural areas. The extent of this contribution, however, is governed by the other competing demands for crop residues which - depending on the crop - include use for animal fodder, building purposes, handicrafts and fertilizer (either through burning or being

TABLE 4.3 Production of Residues from Cereal Crops in Developing Countries (per Hectare)

Crop	Crop Yield[1] tonnes/ha/yr range	average	Crop:Residue[2] Ratio	Residue Production[3] tonnes/ha/yr range	average
Rice	0.7-5.7	(2.5)	1:2	1.4-11.4	(5.0)
Wheat	0.6-3.6	(1.5)	1:1.75	1.1-6.1	(2.6)
Maize	0.5-3.7	(1.7)	1:2.5	1.3-9.3	(4.3)
Sorghum	0.3-3.2	(1.0)	1:2.5	0.8-8	(2.5)
Barley	0.4-3.1	(2.0)	1:1.75	0.7-5.4	(3.5)
Millet	0.5-3.7	(0.6)	1:2	1.0-7.4	(1.2)

Sources : 1. 1979 FAO Production Yearbook. Figures refer to the range between countries of national crop yield averages. The number in parentheses represent average yield figures for the developing countries as a whole.

2. World Bank (1979), "Prospects for Traditional and Non-Conventional Energy Sources in Developing Countries", Table I-4. Figures refer to approximate crop residue ratios at field moisture content.

3. Calculated from Columns 1 and 2. These figures are, at best, approximate since crop:residue ratios are likely to show regional as well as varietal variations. The highest values for residue production are likely to be somewhat exaggerated because high yielding crop varieties tend to have lower crop:residue ratios.

TABLE 4.4 Per Capita Production of Cereal Crop Residues in Selected Developing Countries

COUNTRY	ENERGY CONTENT OF RESIDUES FROM CEREAL CROPS (GJ/capita/yr)
Argentina	25.0
Thailand	9.3
Malawi	8.6
Brazil	8.1
Nepal	7.1
China	6.8
India	5.5
Upper Volta	5.1
Bangladesh	4.3
Ethiopia	3.3
Peru	2.9
Somalia	2.1
Congo	1.1
Zaire	0.9
Developing Countries (Average)	5.6

Source : World Bank (79), "Prospects for Traditional and Non-Conventional Energy Sources in Developing Countries", Table I-4. Figures based on cereal crop production data from 1977 FAO Production Yearbook, and grain/residue ratios as shown in Table 4.3. Energy content of residues assumed to be 13GJ per tonne. Average figure calculated from 1979 FAO Production Yearbook.
Note that cereal residues are not the only farm agricultural residues. Significant amounts of residues from roots and tubers, vegetables, nuts, cotton, sugar cane etc. are produced in some countries.

left to rot in the field). As was illustrated in the work of Briscoe, in Bangladesh, competing uses can be very important and can mean that very little of the crop residues are currently 'wasted'.[16]

The implications of diverting crop residues for use as an energy source are not always clear. In particular, the effects of removing organic matter that would otherwise be returned to the soil are hard to gauge. In poor soils this is likely to have a deleterious impact for several reasons, since organic matter appears to have an important role in preventing erosion, conserving water and nutrients, and maintaining soil structure. In better soils, where fertility is higher to begin with, the damage is probably less severe although it may be cumulative. Because of these problems some authors discourage the use of crop residues for energy purposes altogether.[86] It should be noted, however, that in many developing countries heavy use of crop residues is already an unavoidable reality.

Assessing the true availability of crop residues for energy uses is therefore a complicated and site-specific task. If the present pattern of crop residue usage is to be changed, careful consideration must be given to ensuring that displaced demands are satisfied through other means. Thus, if crop residues are to be used more for energy purposes, care must be taken to provide substitute animal feeds and to recycle or replenish lost nutrients.

A further factor pertaining to use of crop residues for energy is that of seasonality of supply. Depending on how many crops are planted annually, residues will only be available for a limited time during the year. To provide a year-round supply sizeable storage capacity will be required and this can pose problems, particularly in wet climates.

c) Animal Manure

Animal manure represents a considerable energy resource in countries where there are large numbers of domesticated animals. Dung, when dried, has an energy content similar to that of wood. This may be recovered by direct burning, as is commonly the practice, or by anaerobic digestion to produce biogas (see Section 4.3.1). The main advantage of anaerobic digestion is that it retains the many nutrients contained in the manure, allowing them to be returned to the soil.

The amount of manure produced per head of livestock depends very much on their level of health and the ease with which it can be collected varies according to the husbandry methods employed. Typical manure production figures for various animals are shown in Table 4.5 and estimates of the average amount of dung produced in selected countries are shown in Table 4.6, expressed at GJ/capita/year. For the developing countries as a whole, it averages about 5.7 GJ, which is approximately equal to the amount of crop residues produced.

A number of limitations apply to use of animal manure for energy. Firstly, apart from the practical difficulty of collecting dung, some countries have cultural objections to this practice. Secondly, to use animal manure to produce biogas for cooking there is a certain minimum scale for the process. If cattle dung is being used as the feedstock, for example, roughly 4-5 animals are required to provide enough dung for a small family unit. For families with fewer than this number an individual plant would not be justified and the benefits of using biogas could not be fully realised unless families enter into some kind of communal arrangement with their neighbours.

TABLE 4.5 Manure Production by Domesticated Animals

Animal	Manure Production[1] tonnes/head/yr
Cattle, buffalos, camels	1.00
Horses, donkeys, asses	0.75
Pigs	0.3
Sheep, goats	0.15
Chickens, poultry	0.005

Source : World Bank (1979) "Prospects for Traditional and Non-Conventional Energy Sources in Developing Countries", Table I-4.

1. Figuresgiven are on a dry weight basis.

TABLE 4.6 Per Capita Production of Animal Dung in Selected Developing Countries

COUNTRY	ENERGY CONTENT OF ANIMAL DUNG (GJ/capita/year)
Botswana	55
Argentina	40
Somalia	28
Sudan	22
Bolivia	19
Brazil	16
Nepal	13
Peru	7.6
Upper Volta	7.3
India	6.3
Bangladesh	5.1
Thailand	3.9
China	3.4
Malawi	2.8
Zaire	1.1
Congo	1.0
Developing Countries (Average)	5.7

Source : World Bank (79), "Prospects for Traditional and Non-Conventional Energy Sources in Developing Countries", Table I-4. Figures based on livestock data from 1977 FAO Production Yearbook using residue production factors shown in Table 4.5 and assuming an energy content of 15GJ per tonne for dry dung. Average figure calculated from 1979 FAO Production Yearbook.

d) Household Wastes

The wastes produced by households in the form of sewage, food scraps and garbage represents a small but potentially useful energy source under some circumstances. This is particularly true in urban environments where the amounts produced are greater and where, if not disposed of, they can create a major sanitation problem. Most of these wastes can be broken down by anaerobic digestion and this would yield methane for fuel use along with a stabilised and potentially valuable digester effluent. The beneficial impact on public health would also be significant.

e) Food Processing Wastes

Food processing wastes represent a subsidiary biomass resource deriving from agricultural activities. They include residues such as husks and shells from various crops, cannery wastes, and bagasse from sugar processing. Whilst the total quantities produced are only modest, they can make a significant energy contribution to help fuel the process which produces them,with perhaps a little to spare. Using these wastes for energy can also help alleviate the pollution problem that they sometimes create. A rough global estimate of the amount of wastes produced from selected crops grown in the developing countries is shown in Table 4.7. Taken together, their energy content represents about 2.5% of the total energy consumption of the developing countries.[27]

TABLE 4.7 Estimates of the Production of Selected Food Processing Wastes in Developing Countries (1975)

By-Product	Estimated Production 10^6 tonnes/yr	Approximate Total Energy Content 10^6 GJ/year	Present Level of Use for Energy Purposes
Sugarcane Bagasse	110	1060	High
Rice Hulls	55	790	Low
Coconut Husks	13	185	Low
Cotton Husks	6	110	High
Groundnut Shells	6	100	High
Coffee Husks	2	35	Low
Oil-palm Husks	2	35	High
Oil-palm Fibres	3	20	High
		Total = 2330 x 10^6 GJ (Equivalent to 2.5% of total energy use in developing countries)	

Source : FAO (1979) "Energy for World Agriculture", ed. B.A. Stout, Table 77.

4.2.2 ENERGY CROPS

As a way of increasing the biomass resource base a number of options are available for growing plants specifically for energy purposes - as 'energy crops'. The main alternatives at present are as follows :

- Fast growing trees
- Sugar and starch crops (for ethanol production)
- Oil and hydrocarbon crops
- Herbaceous crops
- Aquatic plants

These are discussed individually below.

a) Fast Growing Trees

Wood energy, as discussed earlier, already makes an important contribution to the energy supply in most developing countries. The majority of this wood, however, is collected in an uncontrolled fasion with little regard to maintaining forest productivity over long periods. Not only does this fail to maximise the potential for wood energy, it damages the productive capacity of the environment and threatens fuel supplies to future generations. To reverse this trend and instead to increase the contribution of wood as an energy source, there is an urgent need to introduce improved forestry practices in the developing countries.

The use of fast growing trees has much to offer in this context. Tree species exist which can be grown on a short-rotation basis, giving wood yield many times that obtainable from unmanaged natural forests. In this way, unproductive land can be converted to highly productive energy plantations where wood is harvested as a crop and used for energy purposes.

Research into identifying and developing optimum species for energy uses is presently underway in a number of countries although it is still a relatively new concept and there is much to be learnt. For use as fuelwood species, trees should ideally have the following characteristics[18]:

- rapid growth, even on poor soil
- ability to stabilise and improve the environment
- minimal management requirements
- disease and pest resistance
- ability to survive droughts and other ecological stresses
- coppicing ability (resprouting from the stump when cut)
- high heating value and suitability for multiple uses

Although research is still in its early days a number of tree species have been identified which appear to be well suited for energy purposes. One group that seem particularly promising in this context are the tropical legumes. This highly diverse but poorly studied plant group has the important ability to fix nitrogen from the atmosphere. They are therefore 'self-fertilizing' and have the effect of improving the soil they grow in. The best known example in this group is the species Leucaena leucocephala, known commonly as the giant "Ipil-Ipil". As well as the characteristics mentioned above, this tree also has the advantage of producing high-protein seeds and foliage which makes an excellent animal feed supplement and green fertilizer.[54] Other tropical legume trees showing fast growth include various Acacia, Calliandra and Sesbania species. Of the non-legumes, Eucalyptus and Casuarina are examples of promising tree types.

The productivity of fast growing trees depends very much on their degree of adaptation to local conditions but can be very impressive under favourable circumstances. Many species will produce yields of wood in excess of 20 m^3 per hectare annually, for instance, when grown on reasonably good soil. In exceptional cases annual increments of as much as 50 m^3 per hectare have been reported for Leucaena, Eucalyptus and several other species.[69]

Whilst these figures indicate what can be achieved under optimum conditions, it cannot be assumed that these results can be repeated anywhere. In many cases it is only the less fertile areas that are available for planting energy crops and tree performance under these conditions as well as being generally lower is also much more variable. Matching the species to local soil and climatic conditions is, therefore, one of the most critical tasks in ensuring the success of tree plantation programmes. Often this will require careful field trials since experience to date in this area is still somewhat limited.

One of the advantages of certain species of fast growing trees, however, is their tolerance of poor soils and their ability - once established - to resist droughts, floods, and other environmental stresses. This enables trees to be planted on land which would be unsuitable for field crops, thereby avoiding the food/fuel conflicts that affect some of the other energy crop alternatives.

Fast growing trees can also be very useful components of reforestation programmes. Reforestation is a major priority in many areas of the world but costs tend to be very high because there is no immediate economic return from planting trees. Fast growing trees, however, as well as being effective in stabilising poor soil, preventing erosion and gradually improving the environment, can also be harvested for fuelwood - which in turn can be sold to help finance the project.

For supplying rural energy needs, the planting of fast growing trees on any small patches of unused land is a logical option for increasing the productivity capacity of the environment. These could includes, for example, areas in and around villages, along roadsides and riverbanks, and in between fields. It could either be done on a private basis of as a community activity in the form of '<u>village woodlots</u>' planted on common land. Obviously, the extent to which fuelwood requirements could be met from these sources depends on the type and quality of land available for these uses. Questions of land ownership are equally important, however, and cooperation between members of the community is likely to be crucial (particularly for village woodlots) because of the need to protect young trees from theft, grazing by animals and fires.

In cases where land is scarce, <u>Agroforesty</u> systems have much to offer. Agroforestry involves the intercropping of trees with field crops, a practice which can allow food and fuel to be produced from the same land.[53] The crops much be chosen carefully so that they complement rather than compete with each other. <u>Eucalyptus</u>, for example, is not a good Agroforestry species due to its high moisture demand and its tendency to produce toxic substances. <u>Leucaena</u>, on the other hand, helps to fertilize its surrounding area since it is a legume and its open leaf structure means that it does not shade-out ground crops as much as some other trees.

Agroforesty methods also help overcome the inherent problem with tree growing schemes of the time delay between planting trees and being able to harvest them. Depending on species and growth rates the optimum rotation length can range from four to ten years. If food crops are grown between the trees for the first few years they will provide a subsidiary source of income to the farmer and help reduce the initial burden of investing money and land in tree growing.

Planting trees which can be used for multiple purposes also improves the economics of tree farming.[23] A number of legume trees, for example, produce leaves which are excellent for animal food and their pods can often be used for human consumption. Other trees yield valuable nuts and fruits, and produce wood which can also be used for building purposes.

All of these ideas illustrate the versatility of tree-based energy schemes. This is a tremendous advantage in view of the importance of blending energy production methods with local conditions and local needs. Trees can be successfully grown, for example, on scales ranging from garden plots to large energy plantations suitable for providing wood for charcoal-making operations, electricity generation, industry or liquid fuel production. The techniques used can be highly mechanised or they can be based on simple labour-intensive methods. Organisationally, tree growing can be managed on a plantation basis using hired labour, or it can be performed by small farmers or community groups.

Attempts to introduce tree planting schemes based on fast growing trees have been made in a number of countries over the last twenty years and currently there is a good deal of activity in this area. Many of these projects have multiple objectives and are aimed at providing employment opportunities and environmental and social benefits as well as a source of renewable energy.

In terms of area planted, the most successful fuelwood

programmes to date have been in China and South Korea.[26] Both of these were based on a high level of community participation and this has been a major factor in their success. Progress has also been made in Gujarat State in India and in a few isolated cases elsewhere. There has also been a large number of projects that have failed, however. Often this has been due to sociological and political reasons rather than technical ones. These factors are vitally important in determining the likely success of tree planting schemes and are dealt with in greater depth in Chapter 6.

b) Sugar and Starch Crops

In countries where there are large areas of reasonably fertile land which is presently under-utilised, the potential exists for growing sugar and starch crops for ethanol production. A variety of crops are suitable for this purpose. Of the sugar containing crops, sugar cane and sweet sorghum are the main examples. These show very rapid growth under good conditions using modern agricultural methods (up to 50 and 35 tonnes/ha respectively), and have the advantage that the sugars they contain are directly fermentable to ethanol. They also yield bagasse as a by-product when the sugar is squeezed from the stalks. Bagasse can be used as a fuel for the energy-intensitve ethanol distillation process and this greatly improves the overall energy balance. The main disadvantage of these crops (particularly sugar cane) is that they require good land and adequate water supplies to provide high yields. They are generally not appropriate, therefore, in countries where there is strong pressure on agricultural land.

The primary starch crop of interest is cassava. Whilst a variety of other plants such as sweet potatoes, corn, rice and other cereals, can be converted to ethanol, they are generally more valuable as foodstuffs and are therefore not considered in this discussion. Cassava has the advantage that it is much more tolerant of poor soil and adverse weather conditions than the sugar crops mentioned. It is presently grown as a subsistence crop in many developing countries and does not require high levels of fertilizer or pesticides to give good yields. It can also be harvested throughout most of the year, allowing for more efficient use of conversion facilities than for sugar crops which must be harvested over a relatively short time-space and do not store well. As a feedstock for ethanol production it has the advantage of having a high starch content, giving high overall conversions, although this starch must be first broken down to simple sugars through pre-treatment steps (saccharification) before it can be fermented. A major disadvantage is that it produces no residue equivalent to bagasse, which can be burnt to help fuel the process. This penalises the overall energetics but may be acceptable if other renewable fuels such as wood can be used instead.

The productivity of the various crops mentioned under Brazilian conditions are shown in Table 4.8, along with estimates of ethanol yields.

A further potential feedstock for ethanol production that is available in some countries is surplus molasses from existing sugar production facilities. For every tonne of cane sugar made, approximately 190 litres of molasses is produced as a by-product. This contains between 50-55% fermentable sugars and yields about 280 litres of ethanol per tonne of molasses when fermented. Molasses is primarily used as an animal feed (mainly in the US and Western Europe) but in remote areas it is often wasted because transportation costs are too high. In some cases it even presents a disposal problem. The

TABLE 4.8 Productivity of Ethanol Crops

Crop	crop yield[1] tonnes/ha/yr	ethanol yield litres/tonne	ethanol yield litres/ha/yr
Sugarcane	50[4]	70	3,500
Sweet sorghum	35[2]	86	3,010
Cassava	12	180	2,160
Cassava	20[3]	180	3,600[3]

Source : World Bank (1980) "Ethanol Production from Biomass in the Developing Countries", Table 8.

Notes 1. Based on current average yields in Brazil.

2. Based on one crop per year. Two crops may be possible in some locations.

3. Potential with improved production technology.

4. Yields of 100 tonnes/ha/yr can be obtained with irrigation and high levels of fertilizer.

World Bank estimate that 5 million tonnes of molasses are wasted annually, 15% of the total amount produced.[3] Converting this molasses to ethanol is probably the most immediately feasible and economic approach to ethanol production. Whilst the overall impact of such a scheme would be obviously limited by the relatively small size of the molasses surplus and economic uncertainties due to fluctuations in the world molasses price it may represent a valuable source of liquid fuel for some countries.

c) Oil and Hydrocarbon Crops

A number of plants exist which produce oils and hydrocarbon substances as part of their natural metabolism. With various degrees of processing and/or blending with other fuels, these products can be used for powering internal combustion engines. Although research into producing and using these compounds for fuel is still in its early days this is a concept that may well have significant potential in the developing countries as a simple and economic way of producing liquid fuel from biomass.

The use of vegetable oils as a fuel for diesel engines is the simplest and most readily applicable example in this category.[39] Vegetable oils consist of various types of triglycerides and are presently produced from a wide range of plants including soya bean, sunflower, groundnut, oil palm, coconut, castor and sesame.*

Most vegetable oils can be used directly in diesel engines,

* Production of vegetable oils in the developing countries in 1979 was estimated at 29 x 10^6 tonnes (1979 FAO Production Yearbook).

either pure or blended with diesel fuel. In vehicle tests in a number of countries, some problems have been encountered with these fuels due to the clogging of injection nozzles and the formation of deposits inside the engine. These problems can generally be overcome, however, with minor engine modifications and through correct tuning. Esterification of vegetable oils may be advantageous if they are to be used in the pure form. This is a fairly simple process, involving heating the oil for a few hours at 30-40°C with methanol or ethanol in the presence of acid. The esters so formed have viscosity and distillation properties similar to diesel fuel but if anything burn cleaner and give higher engine efficiencies.

The production of vegetable oils for use as liquid fuels has several advantages for developing countries compared with other liquid fuel options. First of all the process is very simple. Oils can be pressed from the nuts or beans from which they are derived using rudimentary equipment and requiring only minor energy inputs. No water is required for processing and no major pollutants are produced. Indeed, the 'cake' left over from these plants after oil extraction is generally high in protein and is often very useful as an animal feed or fertilizer. Some also produce by-products which can be burnt for fuel or used as building materials.

The climatic tolerance of oil producing plants varies significantly between species but most of them can grow in a wide range of soils providing the moisture and temperature conditions are suitable. Oil yields from these crops vary widely depending on the agricultural methods used, but under good conditions can compare favourably with the other methods of producing liquid fuels from biomass (see Section 5.3). Coconut palm and oil palm are the most prolific oil producers, yielding approximately 1.5 and 5.0 tonnes of oil/ha/yr respectively under good conditions, and as much as 3.3 and 8.4 t/ha/yr in the highest recorded instances.[60] The yields from oil-producing ground crops (sunflower, castor, soya bean, sesame, groundnut etc.) are lower and range from roughly 0.6 t/ha under good conditions to 1.2 t/ha in the best cases - per growing season of 120-160 days.

Perhaps more important than these optimum production figures, however, is the fact that several oil crop species can tolerate soil and climatic conditions under which few other crops can survive. Castor and sesame, for example, are drought resistant (once established) and coconut palms grow well in sandy coastal regions. Whilst yields will inevitably be reduced in such circumstances, using land which is unsuitable for growing food crops lessens the competition between food and fuel production. This contrasts with the situation for some other energy crops, notably sugar cane, for which fertile land is required. Most of the oil producing plants also have the advantage of being well suited for inter-cropping with other species as part of agroforestry systems.[53]

It should be noted that one factor which may limit the use of vegetable oils for fuel is that, depending on prices, it may be more profitable to sell them for their traditional uses - for cooking and other purposes. The current price of palm oil on the world market, for example, is about $620 per tonne. This compares with an ex-refinery price for diesel oil (for which it would substitute) of approximately $275 - $350 per tonne. At present, therefore, the economic viability of using vegetable oils for fuel is likely to be restricted to locations where production costs are low and diesel prices are unusually high due to transportation or other factors.

Apart from plants which yield vegetable oils, a wide range of species are known which produce hydrocarbons or hydrocarbon-like substances. These also have the potential of being used either as fuels or as feedstocks for chemical production. The natural rubber tree Hevea brasiliensis is the best known example in this category Interest has also been revived in producing rubber from the desert shrub guayale as well as a number of other species.

The latex produced from these plants consists of very high molecular weight hydrocarbons. For use as fuel, on the other hand, lower molecular weight compounds are required and preliminary surveys have shown that there are a large number of species that may be potential candidates for this purpose. The genus Euphorbia appears particularly promising. Relatively small-scale tests with the species E. lathyris in California and E. tirucalli in Japan have shown that harvesting these plants whole, yields of 'oil' of 7-15 barrels per acre are possible (approximately 1-2 tonnes/hectare). Further tests are required to confirm these figures for large-scale plantations and to assess their performance under other conditions. The fact that a substantial number of the 8000 or so known species of Euphorbia are capable of growth under semi-arid conditions may be of considerable relevance to developing countries that are prone to droughts. Significant yield improvements may be possible with these species, as well, through breeding, since few of them have been grown commercially in the past. Research is also needed to find the best way of processing and using the oils that they produce since they are a complex mixture of many compounds, certain of which may be more useful as chemical feedstocks.

Other interesting plants in this context include a number of tree species, such as Copaifera and Croton, which occur naturally in the tropical regions and produce oily exhudates with properties similar to diesel fuel. Little work has been done, however, to determine the possibility of growing and harvesting these trees for fuel on a commercial basis.

d) Herbaceous Crops

In some cases, the harvesting of herbaceous crops may be a viable source of biomass energy. A number of tropical and savanna grasses give extremely high growth rates, due to their high degree of adaptation to local conditions. Indeed, in may cases the vigour of these 'weed' species is a major problem preventing the use of land for agricultural purposes. As a consequence, large areas in some countries have been taken over by these grasses following removal of the original forest cover by slash and burn farmers.

Notable examples of such species include Napier Grass (elephant grass) and the species Imperata cylindrica. Whilst very little research has been done to assess what yields would be sustainable if these species were harvested for energy purposes it is likely that their fertilizer, water and agronomic requirements would be less than for more sensitive crops such as sugar cane. They, therefore, could well be a viable energy alternative in cases where land is avialable but cannot be effectively used for agriculture. Because of their high cellulose content, these types of species are not well suited for conversion to ethanol using current techniques. They can be used effectively for charcoal making, however, and could also provide a feedstock for thermochemical conversion to gaseous or liquid fuels.

e) Aquatic Plants

The yield of certain fresh-water and marine plants can be very high under optimum conditions, with annual productivities of 50 dry tonnes per hectare being reported for several species.[55] Potentially these plants could be grown for energy purposes, although as yet this concept has not been tested on a large scale.

A notable example of a rapidly growing aquatic plant is the water hyacinth (Eichhornia crassipes). This plant is in fact notorious for its ability to spread and clog up lakes and waterways and this represents a major problem in many countries. On the other hand, water hyacinth - as well as numerous other water-weeds, algae and sea-weeds - can also be employed for a variety of useful purposes and are harvested for food, animal fodder and fertilizer.

For use as fuel, anaerobic digestion is the most appropriate conversion method owing to the very high moisture content of aquatic plants. Most species can be digested very effectively, yielding a methane-rich gas and a residue which is useful as fertilizer or animal feed but can also be recycled to the growing system (if necessary) to maintain nutrient levels.

One of the main problems with using aquatic plants as energy crops is that they are often difficult to harvest. Also, owing to their bulk, they require fairly large digesters to handle the quantities of biomass produced. There does appear to be scope for this concept, however, particularly if simple, low-cost digesters can be used (perhaps run on a batch basis), and if energy production can be coupled with other functions - such as waste-water treatment or animal feed/fertilizer production.

The use of aquatic plants for waste-water treatment seems a particularly attractive option. Water hyacinth, for example, has the ability to extract many organic and inorganic compounds from water - thereby purifying it - whilst at the same time producing biomass that can be harvested for energy. By combining these functions the economics of both processes can be considerably improved. This concept appears to have distinct potential in the developing countires where pollution from both industrial and urban wastes is a growing problem.

4.3 BIOMASS UTILISATION

To utilise biomass resources for energy purposes a wide variety of techniques are available. The most efficient way to use most forms of biomass, at least those with reasonably low moisture contents, is to burn them directly for heat as near as possible to where they are grown. This approach has a limited application, however, because demand for energy is often remote from biomass production areas and because direct heat is not the only form of energy required. To reduce transportation problems and to increase the versatility of biomass energy it is often desirable to convert biomass to a variety of more convenient fuel forms. These intermediate fuels (such as charcoal, ethanol and methane) can then be used in a range of end-use devices to satisfy the full spectrum of energy needs. The main ways of converting and using biomass energy are shown schematically in Figure 4.1.

In the following sections the main conversion systems and end-use techniques are outlined. They are dealt with under the following headings :

Conversion systems

a) Thermo-chemical conversion processes

1. Pyrolysis/charcoal production
2. Gasification
3. Liquid fuel production
   - Indirect liquefaction
   - Direct liquefaction
4. Ammonia production

b) Biochemical conversion processes

1. Anaerobic digestion to methane
2. Ethanol fermentation

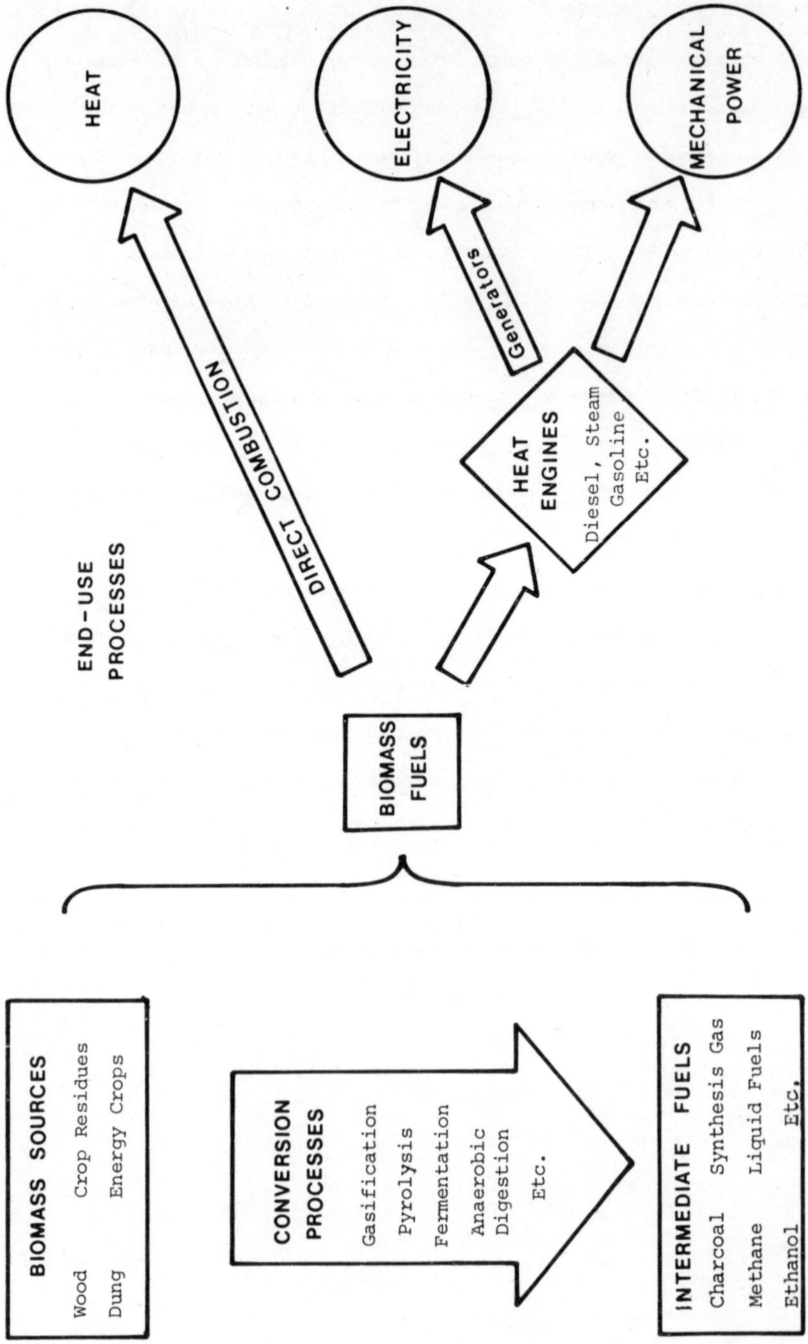

Figure 4.1 Methods of Using Biomass for Energy

End-Use Systems

a) Direct combustion

b) Heat engines

c) Electricity generation

d) Human and animal power

4.3.1 BIOMASS CONVERSION SYSTEMS

a) Thermo-Chemical Processes

Like any other polymer, biomass is unstable at high temperatures and will break down to form smaller and less complex molecules - both liquid and gaseous. Combustion represents a complete oxidation to carbon dioxide (CO_2) and water (H_2O). By controlling the process using a combination of temperature, pressure and various catalysts, and through limiting the oxygen supply, partial breakdown can be achieved to yield a variety of useful fuels. The main thermo-chemical conversion approaches are as follows :

1. Pyrolysis/charcoal production
2. Gasification
3. Biomass liquefaction - direct
   - indirect

These are shown schematically in Figure 4.2 and are described individually below.

1. Pyrolysis/Charcoal Production

Pyrolysis is the thermal decomposition of biomass at elevated temperatures in the absence of oxygen. When biomass is heated in this way a sequence of physical and chemical changes occur which yield a medium energy gas, a complex pyrolysis liquid and solid charcoal in proportions which are determined by the conditions

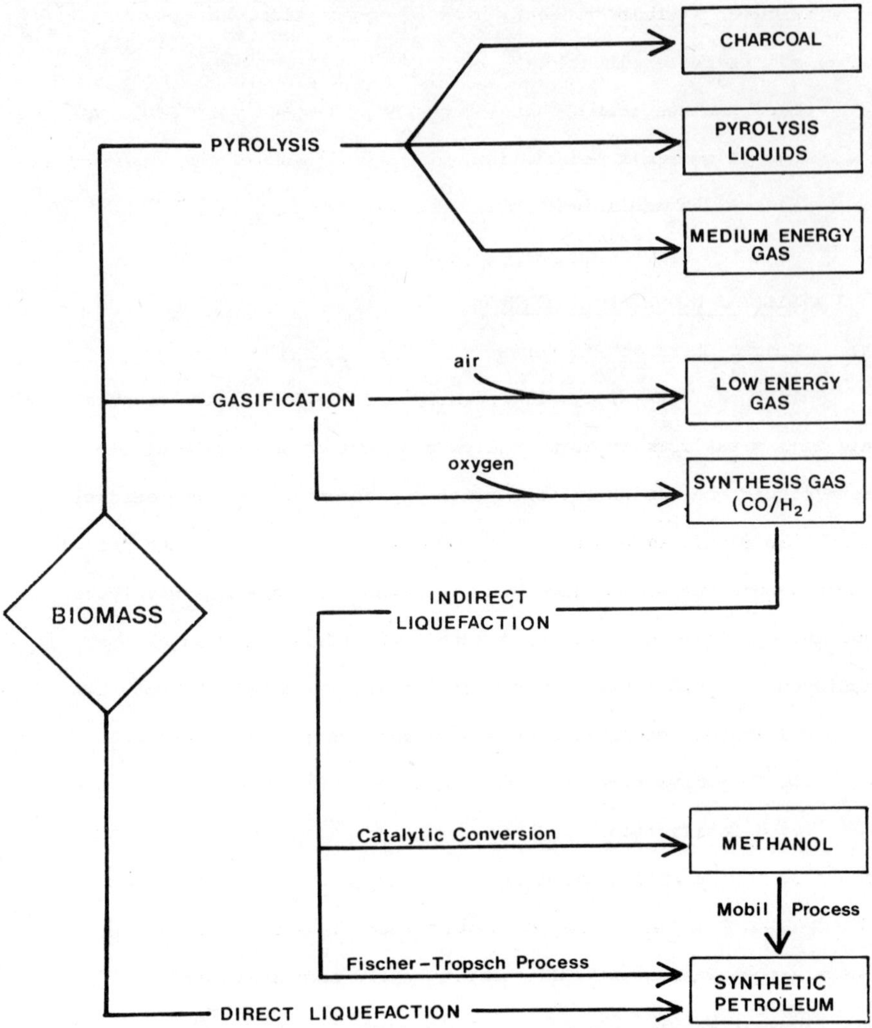

Figure 4.2 Thermo-Chemical Biomass Conversion Processes

(temperature, moisture content, biomass composition, residence, time, etc.).

Pyrolysis is usually carried out at atmospheric pressure and at temperatures between 200°C and 600°C. Under these conditions typical product yields are as follows (weight of product compared to weight of dry feed)[61]:

charcoal	30--35%
organic liquids	18-20%
gas	20%

The liquid products consist of an aqueous solution (commonly known as pyroligneous acid) containing mainly methanol, acetone and acetic acid, and an insoluble fraction of tar and pitch. The gaseous products are largely carbon dioxide, carbon monoxide, methane and hydrogen with an overall heat content in the range of 10,500-15,000 kJ/m^3 (as compared with 37,500 kJ/m^3 for natural gas).

High temperature pyrolysis processes also exist in which wood is heated to temperatures of up to 1100°C. Under these conditions most of the liquid and solid fractions are broken down and gaseous products predominate. The yield of hydrogen is also greater and the gas has a higher energy content. This makes it suitable for use as a 'synthesis gas' for chemical production (see below).

The design of pyrolysis equipment depends very much on the type of product desired (liquid, gaseous or solid charcoal) and the economic constraints on efficiency. For maximum energy efficiency all the end-products must be collected. Unwanted products can then be recycled to provide the necessary heat to maintain reaction temperature. Thus, to produce charcoal efficiently, part or all of the volatile products can be used for drying and heating the feed material. In high temperature pyrolysis, char and liquid products are sacrificed to maximise gas production. Finally, destructive distillation of wood is centred around collecting condensable liquid products at the expense of gaseous and solid components (before the advent of the modern petrochemicals industry this process was the primary source of methanol as well as many other chemicals).

Maximising efficiency, on the other hand, greatly increases the cost and complexity of the process. For production of charcoal - which is by far the largest application of pyrolysis at present - the simplest and least capital-intensive approach is to burn part of the biomass feed to provide the initial heat necessary to begin the carbonisation process. The oxygen supply is then cut off; the subsequent exothermic reaction proceeding by itself. To cut down equipment costs, gaseous and liquid products are allowed to escape leaving only charcoal as the final product.

The most commonly used charcoal producing methods in the developing countries are simple pit kilns and wood piles covered with earth or vegetation. Fairly large pieces of wood are generally required. A fire is started and once it has reached a sufficient stage it is covered and the air supply is cut off. Carbonization takes from days to weeks depending on the size of the pile. Whilst these methods are very inefficient due to the loss of most of the volatile components and because they are difficult to control, they do have the overriding practical advantages of being simple to build and requiring virtually no capital expenditure. Typically, between 8 and 12 tonnes of wood are required to produce one tonne of charcoal using covered-pile methods, while pit kilns are slightly more efficient (requiring approximately 7-8 tonnes of wood). Assuming charcoal has a heat content roughly double that of air dry wood on a weight basis (30 GJ/tonne as opposed to 15 GJ/tonne for air dry wood), the energy efficiency of charcoal production using these methods is only in the order of 17-29%.

Improvements in conversion efficiency can be obtained using slightly more sophisticated charcoal kilns made of brick, concrete or metal. These are easier to control and produce a cleaner end-product. Portable steel kilns, for example, have been used in several countries in Africa for producing charcoal from logging residues.[25] These produce approximately 6 tonnes of charcoal per month and cost in the order of $1000 each (1976), but it has been reported that they have a limited lifetime (2-3 years) due to problems of corrosion and overheating.[40] Brick and masonry kilns last longer and can be cheaper to build if local materials can be used, although they are less mobile. Examples of these include the Brazilian 'beehive' kiln and large 'Missouri' kiln which handles up to 40 tonnes of charcoal per month. Their efficiency again depends on the type of wood and the operating conditions but in general they require between 5 and 7 tonnes of wood per ton of charcoal (corresponding to an energy efficiency of 29-40% for the conversion process).

A number of larger and more sophisticated devices have also been produced which are even more efficient, but they tend to be very expensive. These include various continuous kilns, and retorts which collect the liquid products and recycle the gaseous components. Most of them require a feed material that has been reduced to a fairly small particle size but they are useful for producing charcoal from wastes such as sawdust and bark. One of the most sophisticated pyrolysis devices is the 'Lamboitte' continuous retort, developed in the 1940s. This produces up to 55 tonnes of charcoal per day at a very high efficiency (roughly 57% of the energy in the feed is retained). These plants do, however, cost several million dollars to build. More recently, Georgia Tech have been experimenting with simple and relatively small-scale retorts (producing in the range of 1-6 tonnes of charcoal per day), that might be suitable for use in less developed countries.[20] These operate by burning part of the charcoal producted to maintain reaction temperature and in pilot plant studies have shown yields of 0.25 tonnes of charcoal, 0.15-0.2 tonnes of oil and 0.3-0.35 tonnes of pyrolytic gas per tonne of dry wood feed. Field tests with these units are presently being performed in Indonesia and Papua New Guinea.

A comparison of the efficiency, cost and liftime of some of the main types of charcoal-making equipment is shown in Table 4.9.

As a fuel, charcoal has a number of advantages. It has a high heat content, similar to that of good quality coal. It burns with a hotter flame, with practically no smoke and emits very few polluting substances due to its low sulphur content. For cooking, it is substantially more efficient than burning wood. It can also be pulverised and used in industrial char-oil fuel injection equipment (see Section 4.3.2a) and in cement manufacture. In addition, its properties as a reducing agent make it ideal for smelting metals and for steel making.

Because of its higher energy content per unit weight it is also cheaper to transport than wood. This can be a major factor when demands for urban or industrial uses are far from forest areas. Thus, whereas the inefficiencies of charcoal production generally make it more economic to burn wood in areas near to forests, as transportation distances increase a 'break-even point' is reached beyond which the cost of producing, transporting and using charcoal becomes cheaper.

TABLE 4.9 Characteristics of Charcoal Making Devices

Type	Annual Production (tonnes charcoal)	Initial Investment ($US, 1976)	Life of Device (years)	Wood Consumption[1] per tonne of charcoal (tonnes)	Energy Efficiency[1,2] (percent)
Earth Pile	various	none	one firing	8 - 12	17-25%
Pit Kiln	various	'small'	1-2 years	7 - 8	25-29%
Portable Metal Kiln	72	$1,000	3 years	5 - 7	29-40%
Brick Kiln (Brazil type)	150	$800	5 years	5.7	35%
Concrete Kiln (Missouri type)	310	$2-10,000	7 years	5	40%
Continuous Retort	20,000	$2,000,000	30 years	3.5	57%

Sources : E. Uhart (1976) "The Wood Charcoal Industry in Africa"
A.C. Harris (1978) "Charcoal Production", Proc. 8th World Forestry Congress, Jakarta
D.H. Wood et al (1980) "The Socio-Economic context of Fuelwood use in small Rural Communities"

Notes :
1. These figures are approximate and depend on many factors such as wood type, moisture content and operating conditions.
2. Energy content of charcoal as a percentage of energy content of wood. Charcoal is assumed to have double the energy content of air dry wood.

The position of this break-even point varies depending on the following factors :

- wood production costs
- efficiency and cost of conversion to charcoal
- transportation costs
- selling price of wood compared to charcoal*

These costs vary between countries. In East Africa, for example, Earl reports that the break-even distance based on local conditions in 1970 was 82 kilometers (assuming selling costs of wood and charcoal reflected energy content, that wood is derived from natural forests and that efficient charcoal kilns are used which only require four tonnes of wood per tonne of charcoal).[25] The break-even distance will be substantially greater if (more expensive) plantation-grown wood is used and if charcoal making equipment is less efficient than Earl's optimistic figures. On the other hand, increases in transportation costs due to fuel price rises will tend to reduce the break-even distance.

It must be recognised, however, that these calculations refer to economic costs only. If, as is traditionally the case, charcoal is made using very inefficient methods and without replanting the trees that are cut down, economic costs are a very poor indicator of the true environmental and social costs of the charcoal making process. When the costs in terms of deforestation, erosion and environmental degradation, are taken into account traditional charcoal production methods emerge as being acutely wasteful - and very damaging when

---

* Charcoal is always more expensive than fuelwood on a per kilogram basis. This is due partially to its greater heat content and its increased efficiency as a cooking fuel, but it also reflects various preferences based on convenience, traditions, smokelessness etc.

practised on a large scale. On the other hand, it is equally important to realise that for many of the people who use these traditional methods charcoal making provides their livelihood. For them, there are often very few viable alternatives to these practices and their needs should not be ignored in planning charcoal policies.

The use of liquid products from pyrolysis for fuel and chemicals is not a new idea, although it is not currently practised to any significant degree. As was mentioned earlier, the destructive distillation of wood was used for many years as a source of methanol, acetic acid, acetone, tars, pitch and many other compounds. This practice was largely displaced, however, by the advent of cheap fossil fuels from which these derivatives could more easily be made. Increasing energy prices may well reverse this trend, although there are a number of inherent problems involved in the use of pyrolysis liquids. These include the problems of corrosion (due to the presence of organic acids) and the fact that many of the more valuable components in the liquid fraction are present only in small concentrations - necessitating expensive distillation steps for their extraction.

2. Gasification

Gasification refers to processes which break down biomass completely to yield energy-rich gaseous products.[77] Gasification involves an initial pyrolysis step, as described above. Subsequently, through the admission of limited amounts of oxygen or air to the system, partial combustion of the resulting char and oil components is allowed - which raises the temperature and yields a predominantly gaseous end product. The advantage of gasification techniques is that they enable solid biomass to be converted to a more convenient and versatile fuel form with only a minor loss of energy during the process.

A wide variety of gasification methods are available, ranging in size and sophistication from simple units suitable for running small engines to large capital-intensive plants linked to facilities for the manufacture of liquid fuels and chemicals. One basic distinction that can be made between types of gasification techniques is in the source of oxygen for the process. Air gasification is the simplest conversion method, producing a low-energy gas due to the dilution effect of nitrogen in the air. This gas is well suited for direct heating applications or for use in engines. Oxygen gasification on the other hand uses pure oxygen and produces a higher energy gas composed primarily of carbon monoxide and hydrogen. This is often referred to as "synthesis gas" since, apart from heating and power applications, it can also be used for the synthesis of methanol, ammonia and other chemicals. It is, however, a much more sophisticated and expensive process and requires separate facilities to extract the necessary oxygen from air.

For the majority of applications in developing countries air gasification is the most suitable approach. Although the scale of the process can vary over a wide range, gasifier designs generally fall into three main categories : "up-draught", "down-draught" and "cross-draught" types,depending on the way in which the air is fed into the device.* In each of them the feed material is fed in at the top and as it gradually works its way downwards through the gasifier, it is first dried and is then pyrolysed due to the heat emanating from the hotter zones below. Subsequently, in the combustion and reduction zones, oil and char

---

* Other more sophisticated designs include fluid bed and entrained flow gasifiers, see reference 77.

components are partially oxidised, releasing heat and raising the temperature to roughly 500°C. Any remaining carbon reacts with carbon dioxide and water to yield carbon monoxide and hydrogen. Finally, ash falls through a grate at the bottom. Depending on the design the hot gas is removed from either the side, the top, or the bottom of the unit.

The precise composition of the gas from an air gasifier depends on the type of biomass used, the temperature and rate of reaction. Typically, if wood is used as the feed, the gas composition is as follows[15]:

carbon dioxide	10%
carbon monoxide	20 - 22%
hydrogen	12 - 15%
methane	2 - 3%
nitrogen	50-53%
heat content	≈ 5500 kJ/m^3

If charcoal is used as a feedstock for air gasification the primary reaction can be represented as :

$$2C + \underbrace{O_2 + 4N_2}_{\text{air}} \longrightarrow 2\,CO + 4N_2$$

The gas produced, therefore, is primarily carbon monoxide and nitrogen. This has a slightly lower heat content (roughly 4100 kJ/m^3) and is commonly known as <u>producer gas</u>.

The potential applications for air gasification in developing countries are very wide. For industrial heating purposes, equipment which has been designed to burn oil or natural gas can be easily modified to run on gasifier gas and overall conversion efficiencies can be very close to that of direct combustion (70-80% efficiency) if well designed equipment is used. For large-scale power generation, the gas will work well as a fuel for both gas turbines and steam generators.

Air gasification also has a very interesting potential for small-scale power applications; mobile as well as stationary. With varying degrees of engine modification, gasifier gas can be used as a fuel in both diesel and gasoline engines; although it must be cooled first (to increase its volumetric heat content) and must be free of tar to prevent engine fouling.

The degree of engine modification required to run internal combustion engines on gasifier gas depends very much on the engine design.[28] During World War 2 a large number of both gasoline and diesel powered vehicles were successfully converted to run on the gas produced from small mobile gasifiers, usually fed with charcoal. Whilst certainly not convenient, these devices were a very useful wartime expedient that may well be appropriate in the future.

Engine designs have developed since then, however, and present-day engines are generally less well suited for conversion to gaseous fuels.[15] Spark ignition engines can operate solely on gaseous fuels but require major modification to increase compression ratios. In most cases, a simpler option is to use gasifier gases as a supplementary fuel for diesel engines. By connecting the air intake of the diesel engine to the outlet of the gasifier, when the engine is running gas will be drawn into the combustion chamber - thereby enriching the diesel mixture. Under such conditions, the fuel feed control on the engine automatically reduces the diesel flow. Once the engine is warmed up the diesel fuel supply cuts out completely, allowing the engine to run entirely on gas.[32] If the gas supply is cut off the engine will automatically revert to running on diesel but under normal conditions diesel is only required for starting the engine and warm-up. Overall diesel savings can be as much as 80-85% using this approach.

Because of size and weight limitations with mobile gasifiers and the difficulties of fuel storage and fuel feed, the use of gasifiers for static power applications appears to be a more attractive option at present. Gasifiers need not be as compact and they can be simpler since automatic feed and control is not essential. Several designs have been produced which can be built using relatively cheap materials and which require only modest levels of skills for operation and maintenance.

Whilst good design is likely to be crucial if problems of tarring-up, corrosion and engine-fouling are to be avoided, this concept appears to have much potential - particularly in cases where diesel engines are already in operation and where basic mechanical skills exist. A project is presently underway in Tanzania to test this concept under the auspices of the Small Industries Development Organisation (SIDO) and Twente University of Technology in the Netherlands.[76] A prototype gasifier has been built (costing approximately $2,000) and has been successfully tested, using maize cobs as the biomass feedstock. The aim is to provide power for small-scale maize milling operations but the concept would be equally applicable to many other rural and small industrial uses, provided a suitable biomass source is available.

3) Liquid Fuel Production

A number of thermo-chemical processes exist for converting biomass to liquid fuels. These can be crudely divided into direct liquefaction processes and indirect liquefaction (in which biomass is gasified as a preliminary step). Whilst all these techniques are relatively sophisticated and will generally only be suitable for large-scale conversion facilities, they do represent an important energy option for the future because of the heavy premium that liquid fuels carry.

Indirect Liquefaction

The main examples of indirect liquefaction methods for converting biomass to liquid fuel employ synthesis gas as their feedstock. The majority of these processes were originally developed for the conversion of coal and natural gas to liquid products. Technically, however, they should work equally well using synthesis gas produced from oxygen gasification. Although no commercial-scale biomass liquefaction plants exist at present, the following processes are all at a sufficiently advanced stage of development that they could be implemented in the near future :

- i) methanol production from synthesis gas
- ii) conversion of methanol to synthetic petroleum using the Mobil process
- iii) Fischer-Tropsch production of petroleum from synthesis gas.

For each of them, relatively large-scale oxygen gasifiers would be required. If necessary the ratio of carbon monoxide to hydrogen in the synthesis gas can be adjusted using the 'shift reaction' in which excess carbon monoxide reacts with steam to produce carbon dioxide and hydrogen. The individual processes are described briefly below.

i) Methanol Production from Synthesis Gas

This process is used extensively in the petro-chemicals industry for the production of methanol. Although a variety of methods are feasible, the commonest approach at present is to combine hydrogen and carbon monoxide at high pressure (around 100 atmospheres) at about 300°C, using a CuO-ZnO catalyst. The reaction can be represented thus :

$$\underbrace{CO + 2H_2}_{\text{synthesis gas}} \xrightarrow{\text{catalyst}} \underset{\text{methanol}}{CH_3OH}$$

Methanol yields from wood using this process have been estimated at approximately 360 kg per dry tonne of wood. This corresponds to an energy conversion efficiency in the order of 30-40% (energy in methanol divided by total energy input).[28]

Although methanol has only a modest heat content compared to most petroleum based fuels (19.8 GJ/tonne as opposed to 43.7 GJ/tonne for gasoline), it is versatile and can be burnt in both diesel and gasoline engines. Necessary engine modifications are discussed in Section 4.3.2.

ii) Mobil Process

To avoid the need for engine modification, methanol can be converted to high octane gasoline using the newly developed Mobil process. This process is based on the use of shape selective zeolite catalysts and in pilot-plant studies gives conversion efficiencies in the order of 90%.[48] Plans to construct a plant for the production of 12,500 tonnes of synthetic petroleum per day by this technique are presently underway in New Zealand, using methanol derived from natural gas.

iii) Fischer-Tropsch Process

Hydrocarbon fuels can also be produced catalytically from synthesis gas using the Fischer-Tropsch process. This technique was developed in the 1920s and was used extensively in Germany during World War 2 to produce synthetic fuels. It is presently used in South Africa for coal conversion. One of the problems with the process is the wide variety of products formed, including olefins, alcohols and waxes. Current research is directed at finding catalysts which give a cleaner product mix. An advantage of using biomass as a feedstock for this process, instead of coal, is that its low sulphur content reduces the problems of catalyst poisoning which are a feature of Fischer-Tropsch conversion of coal. Major disadvantages are the high capital cost of equipment, the complexity of the process and the requirement for very large-scale facilities.

Another process for indirect liquefaction of biomass that is still in the research stage, but which may have some potential in the future, is the production of synthetic petroleum from pyrolysis gases. This technique, known as the China Lake Process, employs an advanced 'fast pyrolysis' step which produces a gaseous product containing larger amounts of olefins than standard pyrolysis methods.[14] These can subsequently be polymerised under pressure to form a mixture of higher molecular weight hydrocarbons which can then be refined to yield useful fuels. Overall conversion efficiencies in the order of 22% have been estimated, but further work is required to prove this process on a larger scale.

Direct Liquefaction

Techniques for the direct liquefaction of biomss are also being developed, although they are all still in the research stage.[14] The two main examples are the LBL (Laurence Berkeley Laboratory) and the PERC (Pittsburgh Energy Research Centre) process. Both involve the direct hydrogenation of biomass at high temperatures and very high pressures using carbon monoxide and/or hydrogen as the reducing agent. In the LBL process this reaction is performed on an aqueous cellulose slurry derived from the acid hydrolysis of wood chips.[31] With the PERC process, on the other hand, liquefaction occurs following the blending of dried wood powder with recycled product oil and sodium carbonate catalyst.

In each case an oily liquid is produced which can be fractionated for use as fuel. A considerable amount of work will be required on both processes, however, before their technical and economic feasibility can be determined.

4) Ammonia Production

Although not strictly a fuel, the production of ammonia from biomass through thermo-chemical techniques represents an interesting energy-saving option for developing countries. Ammonia can be produced using nitrogen from air and synthesis gas from the gasification of biomass. The process is typically performed at pressures of 200 atm using $FeO-Fe_2O_3$ catalysts and small amounts of other metal oxides. Mass conversion efficiencies of up to 0.5 tonnes of ammonia per tonne of wood have been reported.

b) Biochemical Conversion Processes

1) Anaerobic Digestion to Methane

The anaerobic digestion of biomass by bacteria is a naturally occurring process and constitutes one of the primary mechanisms of decay in nature. Anaerobic digestion refers to the process of bacterial breakdown of organic matter in oxygen-free environments. The methane produced is a useful fuel and can be used for a variety of purposes. The process as a whole, if harnessed efficiently, represents a valuable means of converting biomass (particularly wet biomass) to a more convenient and versatile fuel form.[7] It also has the benefits of providing a nutrient-rich residue ideal for use as a fertilizer and of performing an important sanitary role when used for human waste disposal.

The mechanism of anaerobic digestion is complex, involving a multiplicity of interacting bacterial species. Biomass is first broken down to simple molecules (largely carbon dioxide, hydrogen and acetate) by a group of degradative bacteria. These are then converted to methane by a second group of 'methanogenic' species. The excess carbon dioxide that is formed in the process dilutes the methane somewhat, resulting in a mixture, termed biogas, that contains between 50% and 80% methane. The thermal efficiency of the process is very high, and more than 90% of the energy content of the degradeable biomass is retained in the methane produced. Of the remainder, a small amount is used in creating bacterial cell mass and the rest is lost as heat.

The digestion process is sensitive to a variety of environmental conditions including temperature, pH, nitrogen levels and rates of feed addition. Temperature plays a particularly important role, the bacteria involved in anaerobic digestion existing in two main categories with characteristically different temperature optima. Thermophilic species operate best at around 45-55°C whilst the slightly slower mesophilic types function between 25°C and 40°C. Outside their preferred ranges conversion rates drop off markedly. Below 10°C, digestion is very slow and this can be a problem in cool climates because the energy required to heat a digester can be significant unless it is efficiently insulated.

The range of biomass materials that can be broken down by anaerobic digestion is fairly broad. Unlike the yeast species used in ethanol fermentation (see below), the degradative bacteria in anaerobic digestion systems have the ability to produce the enzymes necessary for breaking down cellulose and this greatly increases the range of potential feedstocks for the process. The main limitation on digestion arises with substrates which contain large amounts of lignin. As well as being extremely intractable, lignin also tends to tie up the remaining fractions and reduces overall conversion. This can be circumvented by various pretreatment processes, allowing a wider choice of feedstocks, but inevitably involves greater costs.

The biomass feedstock most commonly used for anaerobic digestion is animal manure. This can be converted to methane with a net conversion efficiency of 35% to 50% (energy in methane/energy in feed-stock). Aquatic weeds and low-lignin plant matter can also give high conversions, as can domestic refuse. For efficient digestion the carbon to nitrogen ratio cannot be more than about 35 to 1 so with biomass sources that are primarily carbohydrate nitrogen levels must be made up. This can be conveniently done by mixing in a proportion of animal manure or sewage.

The variety of systems that can be used for anaerobic digestion covers a wide spectrum. They range from basic 'holes in the ground' fitted with some kind of gas collection and removal device, to sophisticated reactors equipped with heating stirring and continuous feed facilities. They can be run on a batch basis, or can be fed in a continuous or semi-continuous fashion. Clearly, the most appropriate design depends on the application. Factors such as the climate, the availability and nature of feedstock material and the end-use for the gas and effluent play an important role in system choice as well as the obvious cost considerations.

As a fuel, biogas has both advantages and disadvantages. It has a relatively high heat content, ranging from 20,000-28,000 kJ/m^3 depending on the percentage of methane it contains (compared to 35,000 kJ/m^3 for natural gas) and can be used efficiently for a variety of purposes. With moderate investment in piping and burner equipment, for example, it provides a convenient clean-burning flame for cooking. It can also be used for most other heating applications and with minor engine modifications can be used as a fuel for diesel engines.

The prime disadvantage of biogas as a fuel is that - as with other gases - it is relatively difficult to store and transport. Removal of carbon dioxide from biogas reduces its volume and increases its heating value but compression of the resulting methane remains an expensive process. The alternative is to use large low pressure storage systems, but this is not cheap either. As a result, there is a strong incentive to design systems in which gas is used as it is produced, thereby cutting down the storage facilities required.

In the developing countries, the use of biogas digesters for providing cooking fuel is the major application of anaerobic digestion at present. Programmes have been launched in a number of countries to popularise this idea, notably in India and China but also to a lesser extent in Thailand, the Phillipines, Kenya, Taiwan and several others. The experiences of the Indian and Chinese programmes illustrate the potential for biogas in developing countries, as well as some of the problems.

In India, approximately 80,000 household biogas plants have been built, mostly to a design promoted by the Khadi and Village Industries Commission (KVIC).[51] Constructed of brick, these use cow dung ("gobar") diluted 1:1 with water as feedstock. They are unstirred and unheated and have an output of 1.7-10 m^3 of gas per day depending on the size, although temperatures in winter can reduce this amount. Gas is collected in an inverted steel drum which floats on the slurry. As fresh material is added, digested slurry is displaced through an outlet pipe. When dried this effluent provides a rich organic fertilizer.

A number of problems have been encountered in attempts to introduce these units. They are expensive to build, costing in the order of $300 for a plant producing 2 m^3 of gas per day (1977) - which is about as much as the average rural household earns in a year. Furthermore, corrosion of the steel gasholder has caused difficulties since these cannot easily be repaired locally and are costly to replace. To help get round these problems a new type of design has been developed known as the 'Janata' biogas plant. This has a fixed roof and is based on Chinese designs (which are described below). Its main advantages are its reduced cost (40-70% of similarly sized KVIC plants) and the fact that by replacing the steel gasholder with a fixed roof if cuts down maintenance problems as well as heat loss during cold periods.

A major factor which limits the impact of household biogas units in an Indian context, irrespective of their capital costs, is that very few of the rural population have enough animals to justify building their own plant. Roughly 4-5 cattle are required to provide enough dung to run even a small unit and only about 10-12% of farmers in India possess this many.[1] Inevitably, therefore, the spread of household biogas plants will be limited to the wealthier members of the community.

Recently there has been significant interest in the possibility of building community biogas plants in India. These could take advantage of the considerable economies of scale in plant construction costs. It is estimated, for example, that the capital cost of a KVIC type plant producing 60 m^3 of gas per day is only 37% of that of a 2 m^3 per day plant - per unit volume of gas produced. A community plant would also allow the pooling of all the animal dung produced in the village and would make the use of human wastes easier if units could be coupled to public latrines. If sufficiently large, plants would justify hiring trained operators to look after them and ensure that they are functioning efficiently. Finally, any excess gas - over and above that needed for cooking - could be used for productive uses such as pumping drinking water, electricity generation, irrigation, grain milling and small industry. Some of these activities could be run on a commercial basis and would help to finance the initial investment and running costs of the plant.

For a community biogas plant to be successful, obviously one of the primary requirements is to devise an equitable system for the collection of dung and for the distribution of the gas and effluent produced. This could be done on an exchange basis, or by assigning a monetary value to each of these inputs and outputs and buying and selling them. Alternatively, products could be simply allocated in accordance with community needs. Clearly, a considerable degree of cooperation within the village will be required for any of these systems to work smoothly.

Experience to date in the use of large-scale communal biogas plants in India has been fairly limited but has highlighted many of the problems that are likely to arise. For example, based on preliminary observations from a community plant built in the village of Fatehsingh-ka-Purwa in Etwah District of Uttar Pradesh in 1978,[68] the following points were apparent :

1. Well designed community scale biogas plants can be operated in the context of an Indian village, although gas production was rather less than was predicted.
2. It is unlikely that the capital cost of community plants could be met by villagers. An initial subsidy is therefore essential.
3. A source of income must be provided from the plant, in some way or another, so as to pay for running costs.
4. Poorer members of the community are unwilling to pay for biogas produced from communal plants, as they have always previously collected fuel for free.
5. Factionalism is a problem and it is difficult to get everyone to cooperate and contribute their dung.

The social and economic problems of introducing community biogas schemes are discussed in greater depth in Section 6.

Success in implementing biogas schemes has been more marked in <u>China</u>.[85] Since 1972, as many as 7 million biogas plants have been built, 70% of them in the densely populated and sparsely forested province of Szechuan where a quarter of the population now use biogas. Most of these are family size units, although some larger scale facilities have also been built to handle sewage from schools, hospitals, factories and the like.

A typical Chinese digester is designed for a family of five, having a capacity of around 10 m^3 and producing 1 to 2 m^3 of biogas per day. The designs used are very simple and consist of an underground chamber which serves as both the digester and the gas collector. Gas collects in the top of the chamber and pressure is relieved by the displacement of water into an upper compartment. Pig manure is the primary feedstock used but human wastes and crop residues are also added in many cases (the recycling of human wastes for use as fertilizer is a long-standing tradition in China, so, unlike many other countries, there is little resistance to its use in biogas plants).

Because these plants are built using locally available materials and because labour is supplied free of charge by the family or community, Chinese biogas plants are substantially cheaper than either the KVIC or Janata plants in India. Material costs as low as 30-60 Yuan (18-36 US dollars) per plant have been reported for household size units, which represents between a half and one month's income for a rural family.[68] Since many families in China also own a few pigs and because these are often kept penned, which increases the ease of dung collection, the ownership of biogas plants is open to a much wider section of the population than in India.

One disadvantage of the Chinese designs is that gas evolution is not as high as with other systems owing to the problems of ensuring airtightness with the simple construction materials used. Also, because of the limited degree of agitation, a scum tends to develop on top of the slurry which must be removed twice a year, along with the sludge that settle to the bottom due to the use of crop residues as a feed material. This low production of gas and the inconvenience of cleaning out digesters, has not posed a major barrier to their introduction, however. This is partly a result of the fairly modest fuel requirements of Chinese families due to their quick cooking methods and partly due to the greater general awareness of the hygiene benefits of biogas digesters and the usefulness of the sludge as a fertilizer.

An important limitation to the expansion of the Chinese biogas programme is that in many regions, particularly in the North, winter temperatures are too low for the digestion process to proceed and this means that biogas plants can only be operated for part of the year.[79]

In summary, the use of biogas digesters for providing cooking fuel has many theoretical advantages. Its applicability in practice, however, is conditional on having sufficient financial resources to be able to afford the plants, adequate dung or other feedstock (and water) to run them and the necessary technical skills and motivation to build and operate them. Each of these areas provides problems in most developing countries although research into methods of reducing digester costs and improving their performance can obviously help.

In some developing countries, the immediate potential for anaerobic digestion systems is greater in other applications, such as <u>hygiene and pollution control</u>. Anaerobic digesters are used extensively for sewage treatment in some developed countries. They are also useful for the disposal of dung from piggeries, cattle feedlots and many industrial operations that produce organic wastes - such as breweries, food processing factories, distilleries, etc. In these applications it is often the waste-disposal aspect of the plant that provides the primary incentive for building them - the gas and nutrient rich effluent representing subsidiary benefits.

There is also potential for using anaerobic digesters as part of various integrated systems for the production of food, energy and fertilizer.[28] Possible options include schemes in which aquatic weeds or algae are grown on ponds and harvested for feeding pigs. The dung from the pigs can be digested anaerobically to yield gas and the nutrient-rich residue can be recycled to the pond - where fish can also be grown. Alternatively, aquatic weeds and various herbaceous plants can be grown as crops, digested to produce methane, and the residue used for animal feed. A large number of theoretically attractive permutations on this basic theme have been proposed, but as yet field experience in developing countries is relatively limited.

2) Ethanol Fermentation

The production of ethanol from fermentable sugars using yeast is an old and well understood process although the concept of using the ethanol for fuel is a relatively new one. As described earlier, a variety of crops can be used as feedstocks for this process, the main examples being sugar cane, sweet sorghum, cassava and the various cereal crops.

By far the largest ethanol programme today is in Brazil.[82] Ethanol production in Brazil has been heavily promoted since 1973 as a way of reducing the country's severe balance of payments problems that resulted from oil price rises. To date the programme has been based primarily on diverting sugar cane from sugar production to use for ethanol manufacture. For the future, however, plans are underway to open up new areas of virgin land and to diversify the feedstock to include sweet sorghum, cassava and babassu palm fruit.

Present production of ethanol in Brazil amounts to approximately 5 million cubic metres annually, which allows all of the gasoline used in the country to be blended with 20% ethanol. There are now 200,000 cars in the country which have been built or converted so that they can run on 100% ethanol, and strong price subsidies and other incentives are offered to encourage ethanol use.

A substantial ethanol programme has also been started in Zimbabwe where gasoline is presently mixed with ethanol to a level of 15%.

The process of ethanol fermentation involves the conversion of simple sugars to ethanol and carbon dioxide by yeasts. Biochemically, it is highly efficient and virtually all of the energy in the sugar is retained in the ethanol produced.

Removal of ethanol from the fermentation broth is an energy-intensive step, however, and is hampered by the fact that the maximum concentration of ethanol that can be achieved in the broth is only in the order of 10% to 12%, because high concentrations of ethanol are toxic to the yeast. Ethanol removal is presently performed using distillation techniques but work is underway to develop alternative extraction processes with a lower energy requirement. Improved heat exchangers, vacuum distillation, and the use of water absorbants and membrane techniques are possible alternatives in this context.

When sugar crops such as sweet sorghum and sugar cane are used for ethanol production, sugary juices can be squeezed from the plant and fermented directly. The bagasse left over from juice extraction can then be burnt to fuel the subsequent distillation steps.

Starch crops such as cassava and cereal crops can also be used, although the starch must be broken down to simple sugars (monosaccharides and disaccharides) before it can be fermented, using the saccharification process. This involves mixing the substrate with water, heating it, then subjecting it to enzymic hydrolysis. Starch crops produce no by product that is equivalent to sugar cane bagasse, however, so energy must be supplied from an external source to fuel the distillation process. The use of other renewable fuels (such as wood from a fast growing tree plantation) is the logical solution to this problem since if oil was used the amount required would often be nearly as great as the quantity of ethanol produced. Estimates of the energetics of ethanol production from sugar cane and cassava are shown in Table 4.10.

The disposal of stillage wastes from ethanol plants presents a problem in developing countries that deserves much fuller attention than it is presently receiving. Stillage is the liquid effluent from the distillation system and is produced in quantities which are 10-13 times the volume of ethanol produced. Whilst it does contain valuable nutrients it is also very dilute, making it difficult to use in a cost-effective fashion. Two beneficial uses that it could be put to are

TABLE 4.10 Energy Analysis of Ethanol Production from Sugar Cane and Cassava

	Sugar Cane	Cassava with Tree Farm	Cassava without Tree Farm
ENERGY INPUTS			
Farm System :			
Fuel	0.07	0.10	0.07
Fertilizer/chemicals	0.04	0.02	0.01
Sub-total	0.11	0.12	0.08
Ethanol Plant :			
Fuel	1.26	0.84	0.84*
Chemicals	0.01	0.01	0.01
Sub-total	1.27	0.85	0.85
Total Energy Inputs	1.38	0.97	0.93
ENERGY OUTPUTS FROM BY-PRODUCTS			
Bagasse	1.26	-	-
Wood	-	0.84	-
NET ENERGY INPUT	0.12	0.13	0.93
ENERGY OUTPUT IN ETHANOL	1.00	1.00	1.00

- All figures in GJ -

* Fuel from external source

Source : World Bank (1980) "Alcohol Production from Biomass in the Developing Countries", Table 10. Data taken from Brazilian case study.

for animal feed or fertilizer. The former use requires an evaporation step to concentrate the liquid, making the resultant feed relatively costly unless solar evaporation is used. Use of stillage for fertilizer, on the other hand, requires a distribution network for spreading it on the land. Although both these practices are ecologically sound their economic cost creates practical limitations. Dumping of stillage wastes into rivers is the commonly used alternative and this practice is creating severe pollution problems wherever it occurs.

Another alternative for stillage disposal that is presently being investigated involves growing water hyacinths on the effluent. These plants are able to extract nutrients from water thereby purifying it. The water hyacinth can then be harvested for use as animal feed or fertilizer, or it can be digested anaerobically to yield biogas and a concentrated nutrient-rich residue.

To increase the overall potential of ethanol production a substantial research effort is underway to develop systems that will allow a wider variety of biomass feedstocks to be used. In particular, a number of schemes have been put forward for the conversion of cellulosic and ligno-cellulosic materials to fermentable sugars.[12] These feedstocks have the advantage that their use for energy purposes does not compete with their use for food. Prices are often cheaper, therefore, and the food-fuel conflict is reduced. Examples of feedstocks that could potentially be used in this way include wood, crop residues, and herbaceous crops.

The biochemical problems of using these types of biomass for ethanol production are three-fold. Firstly, the potentially useable cellulose and hemicellulose components of the plant are generally tightly bound up by the lignin fraction. Secondly, once freed, they must be hydrolysed to smaller sugars before they can be fermented. Thirdly, the five carbon sugars (pentoses) produced from hemicellulose breakdown cannot be used by yeast for ethanol synthesis. This reduces overall conversion levels significantly because many of the potential feedstocks contain as much as one third hemicellulose.

To solve the first of these problems a variety of pretreatments can be used, including exposure to strong acids (with or without heat) and solvent separation. To hydrolyse the cellulose and hemicellulose fractions, either enzymes or weak acids can be used. Some schemes involve the use of cellulase enzymes produced separately by specially selected fungal strains of the *Trichoderma* family. Others employ in-situ bacterial hydrolysis, using thermophilic bacteria in some cases to speed up rates and reduce contamination problems. Once the cellulose has been broken down to simple sugars (hexoses) it can be fermented to ethanol either with yeast or various bacterial species. To use the pentose sugars generated from hemicellulose a variety of bacterial strains have been isolated, some of which produce other types of alcohols (such as butanol or butenediol) which are also useable as fuels, or chemicals.

Overall, however, although many of these options have important theoretical advantages they are still very much in the research stage. Even if growth rates and conversion efficiencies do improve, they are unlikely to be very cheap because of the levels of sophistication required for the processes.

4.3.2 END-USE SYSTEMS

a) Direct Combustion

The simplest and most direct method of recovering energy from biomass in the form of heat is by the traditional method of complete combustion, i.e. burning in the presence of air until all of the organic material has been converted to carbon dioxide and water. The burning process can take place under a variety of conditions and in a variety of furnaces, ranging from an open fire which is the least efficient, to a fluidised bed combustion unit which is the most efficient.

The combustion process begins after residual moisture in the fuel has been driven off and the temperature has been raised to the point where thermal decomposition occurs (200-400°C). Large volumes of gaseous and liquid products are given off, part of which are combusted in the gas phase. The remainder escapes unburnt, carrying with it small particles of combusted and semi-combusted material which are visible as smoke. A residue of solid carbon is generally left and this burns more slowly, but with a greater heat output.

The efficiency of combustion devices in terms of useful heat extracted per unit weight of fuel is determined by three main factors :

- the moisture content of the fuel
- the completeness of combustion
- the efficiency with which heat is transferred to the desired sink (boiler tubes, cooking pot, etc.)

Moisture content is particularly important and affects the heat available from burning biomass in several ways. Firstly, on a simple weight basis a kilogram of wet fuel will contain much less combustible material than the same weight of dry fuel. Secondly, the energy released on burning will be reduced by the amount needed to evaporate any moisture that is present (approximately 2.2 GJ/tonne of water). Finally, the efficiency of most combustion devices is reduced by up to 10-20% when moist biomass is used - due to the lowering of the flame temperature and the resulting increase in emission of unburnt particulates (unless specially designed furnaces are used). The combined effect of these factors means that a tonne of air dry wood at 25% moisture content will yield approximately twice as much useful heat, when burnt, as the same weight of freshly cut wood (50% moisture). On a volume basis, the useful heat that can be derived from a cubic metre of freshly cut wood can be increased by a factor of as much as 1.5 by air drying and by 1.7 if it is completely oven dried before burning.[28]

Cooking Stoves

The commonest form of biomass combustion systems are the open fires and simple stoves used throughout the developing world for cooking and heating. Compared to modern furnaces these devices are extremely inefficient, although they do have the major practical advantage of being very cheap as well as being familiar and convenient to use.

Because of the perceived inefficiency of present cooking methods and the growing awareness of domestic fuel problems, a considerable effort has been made in a number of countries in the last 20-30 years to design and introduce improved stoves. Apart from the problem of having to use wet fuel at times, especially during the rainy season when drying is difficult, inefficiencies of traditional stoves and mostly related to the following factors :

- incomplete combustion of fuel and 'smokiness' due

to inadequate and uneven air flow

- poor heat transfer to cooking pots and loss of heat due to convection of hot gases, conduction and radiation to the surrounding area
- loss of heat during the heating up and cooling down process when no cooking is actually performed.

Significant increases in efficiency can be made but there are a number of inherent problems in designing improved stoves, particularly those related to defining and measuring stove efficiency. Stoves are used for many purposes during the cooking process - simmering, boiling, frying and baking - as well as for room heating and various other subsidiary functions, such as keeping away insects. They are also used with different fuels at different times of the year and significant variations in fuel consumption can occur with different operators depending on the care taken in controlling the fire and conserving fuel. Defining a simple test for stove efficiency which is repeatable and is representative of all the stove's uses is, therefore, very difficult.[10, 24]

Accepting that the definition of stove efficiency is far from perfect, the efficiency of open fires and traditional stoves is generally quoted as being in the order of 5-10% (percentage of energy in fuel that is transferred to the food).[45, 19] Charcoal burning stoves have a higher efficiency, probably 10-20%.

Efforts to improve stove efficiency have, for the most part, centred on enclosing the fire, regulating the flow of air into the stove and in some models, providing chimneys and adding thermal mass for heat storage. For enclosed stoves, convering of pot holes when not in use and ensuring that cooking pots fit these holes is also important. Because of the need to keep costs down, the ability to use cheap and locally available materials and simple construction methods is obviously vital. Likewise, because cooking methods and stove requirements vary very much between different parts of the world, it is important that stoves are matched to local needs. This point is highlighted when one considers the difference in cooking styles between India - where slow simmering is employed - and China, where quick frying is the predominant cooking method. Similarly, if mobility or room heating are an essential feature of present stoves this must be taken into account in designing improved stoves.

A variety of 'improved' stoves have been produced based on these principles. Examples include the 'Lorena' stove designed in Guatemala, which are large fixed devices made from a hardened sand and clay mixture and the Indian 'smokeless chula' made from mud. Other more mobile stoves are made from simple metal canisters lined with ceramic material.

Under experimental conditions, and for fairly closely defined tasks - such as boiling water - impressive fuel savings can be realised using these stoves. Efficiency can be doubled or tripled compared to open fires, allowing fuel consumption to be reduced by half or more. Fairly minor modifications to existing stoves can also give useful efficiency improvements.

If these fuel savings could be translated into practice, under field conditions, they would obviously have a major impact on reducing the pressure on cooking fuel supplies. As yet, however, major difficulties have been encountered in introducing new stove designs even though they may be technically superior to traditional types.[71,43] Many social and economic considerations are involved in this technology diffusion process - factors such as convenience, traditions and lack of perception of the need to conserve fuel. Furthermore, even if stoves are accepted - and used efficiently - there is no guarantee that this will automatically reduce fuel consumption levels by the amount predicted from the increase in stove efficiency. Users may prefer to burn the same amount of fuel and simply use more heat.

Overall, therefore, while the theoretical potential for decreasing fuel consumption using more efficient stoves is very great, this may be a difficult goal to achieve in practice. This question is discussed in more detail in later sections.

Large-Scale Combustion Systems

For larger scale applications, a wide variety of direct combustion systems are available with various levels of sophistication. The main types of combustion systems can be divided into four broad groups.[14] These are listed in order of increasing sophistication as follows :

- pile burning
- grate burning
- suspension burning
- fluidized bed combustion

The main characteristics of these types of combustion systems and their variants are outlined in Table 4.11. They each have their own advantages and disadvantages and choice of the best system for a particular application obviously depends on the relative importance of many system features. These include factors such as :

i) Biomass cost and the resulting importance of combustion efficiency in determining overall costs.

ii) The type of biomass feedstock that is available and the need for drying and reduction of particle size.

iii) Scale of facility required and capital investment limits.

TABLE 4.11 Characteristics of Biomass Combustion Systems

Type of Combustion System	Main Variants	Main Characteristics
Pile Burning	Dutch Ovens Cyclonic Furnaces	Biomass is burnt in a pile within the furnace. Dutch ovens are the oldest and simplest type of furnace and can take large pieces of biomass but are difficult to control. Cyclonic furnaces use a jet of circulating air to improve conversion efficiency. Both can handle fairly wet biomass and are generally suited to small to medium scale applications.
Grate Burning	Fixed Grate Moving Grate	Grates are used to separate ash from incoming biomass. Various grate designs exist and are designed to improve the air supply, the uniformity of biomass feed and ash removal.
Suspension Burning	-	Biomass is ground into fine particles and suspended in preheated air before entering the furnace where it is burnt under turbulent conditions. No grate is needed. Biomass must contain no more than 14% moisture.
Fluidized Bed Combustion	-	Biomass is ground and burnt in a floating, agitated 'bed' of non-combustible mineral. This is the most sophisticated and expensive approach but it gives high efficiency and low pollution levels.

Source : D. Bond (1980), State of the Art of Thermochemical Processes for Conversion of Biomass to Energy.

iv) Form of heat energy required (high or low temperature heat, steam, etc.) and whether it is needed intermittently or continuously.

v) Level of sophistication and the consequent need for skilled operators.

vi) Pollution problems.

Well designed biomass combustion equipment can give overall combustion efficiencies approaching that of coal and oil burning facilities (75%-80%). Somewhat larger equipment is required, however, and feed handling costs tend to be higher, particularly if wood is being used (as opposed to, say, charcoal - which approximates very closely to using coal).

An additional and somewhat specialised option for direct combustion of biomass is to use '<u>Biomass char-oil mixtures</u>' in oil burning installations. The concept involves mixing pulverised charcoal with normal fuel oil in such a way that the resulting suspension behaves like a homogeneous fluid and can be burnt in modified boilers and combusters that presently use No.6 fuel oil. This allows partial substitution of petroleum products by renewable fuels to take place without major replacement of existing equipment. The technology is still in the research stage but appears to be promising based on experience with coal-oil mixtures, which can be used successfully in mixtures containing as much as 50% coal.

b) <u>Heat Engines</u>

Heat engines are devices that convert heat energy to mechanical energy. The invention of heat engines and their subsequent widespread adoption, was a crucial component of the Industrial Revolution in Europe. For the third world they are likely to have an equally important role in development.

Heat engines can be divided into two main classes depending on how heat is supplied to the engine. In 'internal' combustion engines heat is generated by the combustion of fuel within the engine, whereas in 'external' combustion devices heat is transferred to the engine from an external source. Within each group a variety of basic designs have been developed and the main types are listed below :

INTERNAL COMBUSTION ENGINES

- Diesel engine
- Otto cycle (gasoline engine)
- Brayton (gas turbine engine)

EXTERNAL COMBUSTION ENGINES

- Rankine (steam engine) - reciprocating/turbine
- Stirling (air engine)

Typical efficiency of these various engine types as well as their capacity range and relative capital costs are shown in Table 4.12.

Biomass fuels can be used in all types of heat engines. External combustion engines have the advantage that biomass can be burnt directly to generate power. With internal combustion devices, on the other hand, biomass must first be converted to compatible liquid or gaseous fuels

TABLE 4.12 Characteristics of Heat Engines

Engine Type	Typical Efficiency[1]	Capacity Range[2]	Capital Cost
Internal Combustion Engines			
Diesel	35-40%	small, medium, large	moderate
Otto (gasoline)	25-30%	small, medium	low
Brayton (gas turbine)	25-30%	medium, large	high
External Combustion Engines			
Rankine - open cycle	10%	medium	moderate
Rankine - closed cycle	20-30%	large	high
Stirling	35-40%	small, medium, large	high

1. Source : "Alternative Automobile Engines", D.G. Wilson, Scientific American, Vol.239, No.1, 1978.

2. Capacity range : small = less than 10 kW; medium = 10-100 kW; large = more than 100 kW

and this inevitably involves significant energy losses in the conversion process. Overall conversion efficiency of the energy in biomass to mechanical power is not the only factor of importance however. Scale options and capital costs are also relevant to the choice of engine type. In practical terms the availability of the necessary hardware and the ease with which presently existing equipment can be modified to run on biomass fuels are additional factors that must be considered.

For small-scale mechanical energy demands internal combustion engines have many advantages. They are generally cheaper to produce than external combustion devices and small capacity units can be made which are both light and efficient. Indeed, it is for these reasons that the diesel and gasoline engines are now the dominant sources of mobile shaft power. They also have the significant advantage of being mass-produced, 'off-the-shelf' technologies. Converting these types of engines to run on biomass fuels, therefore, is the obvious immediate application of biomass energy for small-scale power demands. The range of biomass derived fuels that can be used in diesel and gasoline engines is shown in Table 4.13 along with a listing of the degree of engine modification required.

To meet small-scale mobile power requirements the use of liquid fuels has obvious advantages and possible fuel options include ethanol, methanol, synthetic petroleums and various vegetable oils. There are problems relating to the production of all of these fuels, however, and an alternative that may be attractive for some applications is to use mobile gasifiers to convert charcoal to producer gas and to feed this into internal combustion engines. As mentioned, this approach is not new and was used in a number of countries during the Second World War. It could presumably be revived fairly easily.

TABLE 4.13 Fuel Alternatives for Gasoline and Diesel Engines

Type of Engine	Alternative Fuel	Percentage Substitution of Gasoline/Diesel	Engine Modifications Required
Diesel Engines	methane gasifier gases	80%	Minor[1]
	ethanol methanol	20-45%	Minor[2]
	vegetable oils	20-100%[5]	Minor
Gasoline Engines	methane gasifier gases	100%	Major[3]
	methanol ethanol	20%	None
	methanol ethanol	100%	Major[4]
	synthetic petroleum	100%	None

Notes : 1. Gases are aspirated into the cylinder and ignited by the simultaneous injection of diesel fuel.

2. Alcohols are not sufficiently soluble in diesel fuel to be blended directly. Instead they can be injected simultaneously or fumigated into the air intake. Alcohol containing up to 20% water can be used.

3. Major engine redesign is needed to increase compression ratios.

4. Engines require higher compression ratios, modified carburettors, and measures to avoid corrosion.

5. Purification and esterification are required if vegetable oils are to be used pure.

Source: Office of Technology Assessment (1980) "Energy from Biological Processes", Vol.2.

Technically, due to the start-up problems associated with this kind of system, it would probably be best suited for applications such as trucks and buses in which the vehicle is in use for prolonged periods.

For small-scale stationary power requirements the choice of potential fuel and engine options is wider than for mobile power applications. Engine size is a less important criterion and a greater variety of fuels can be accommodated. Thus, the powering of diesel and gasoline engines from the gases produced by biogas digesters and small gasifiers becomes an attractive option.

In the long-term, the use of small Stirling engines has a number of advantages. These are external combustion devices and as such can burn biomass directly. They also have the potential of using solar energy as a heat source and because of their simplicity, should be more reliable than gasoline and diesel engines. Further research is needed in this area, however, and these will have to be implemented from scratch.

For medium and large-scale power requirements the advantages of internal combustion engines are less marked. Biomass can be efficiently converted to mechanical energy using various types of closed-cycle steam engines and these are the most common technologies used today. Large Stirling engines may have some advantages in the future but are still in the development stages. Gas turbine engines coupled to gasifiers are an additional option for large scale power applications. The trade-offs between using these, as opposed to steam engines, have not yet been fully analysed, however.

c) Human and Animal Power

Human beings and animals can also be regarded as energy conversion devices since they generate mechanical power through the biochemical oxidation of biomass. In many rural situations, power provided in this way is the major source of mechanical energy. In energy terms, the process is rather inefficient since most of the energy in food is expended in maintenance of metabolism. Makhijani and Poole[47] estimate that the overall efficiency of draught cattle generating half a horsepower for 1000 hours per year is 5%, based on the energy content of the food they eat (minus the energy contained in their dung). Revelle makes a similar calculation for human energy conversion and arrives at an overall figure of 3% energetic efficiency for human activities in a typical rural third world village.[66]

When viewed in energy terms draught animals are significantly inferior to mechanical devices. They do have some obvious advantages, however. Their capital cost is low, they run on indigenously produced fuel (much of which is difficult to use otherwise, except for burning), they produce valuable by-products such as milk and manure and they can be eaten. A major disadvantage is that they cannot be switched off when they are not needed. Thus, maintaining animals during periods of the year in which their labours are not required can be a major strain on the agricultural system.

Chapter 5

Overall System Options

The main methods of growing and using biomass for energy were introduced, in general terms, in the previous chapter. Clearly the range of choices of overall biomass systems (production, conversion and end-use) is very wide; and is ever increasing as new processes are developed and old ones adapted. In this chapter an attempt is made to focus in on the types of biomass energy approaches that are most applicable to the needs and conditions of the developing countries.

To this end, a number of specific biomass options are considered for each of the energy need categories identified earlier (domestic needs, rural development and petroleum replacement). Whilst not intended to be in any way definitive, the selection of systems for inclusion in this section is aimed at highlighting some of the options that appear to have the greatest immediate potential and widest applicability. Emphasis is placed primarily on their technical feasibility, although some comments are also made concerning financial costs. No detailed attempt is made to compare these options, however, due to the lack of relevant implementation data and because of the highly site-specific nature of many of the technical and economic variables involved. Likewise, the many socio-economic and other factors that determine what is feasible (or desirable) in practice are omitted and are discussed in subsequent chapters.

5.1 Domestic Energy Needs

The causes and symptoms of the growing domestic fuel shortages in developing countries were discussed in Section 3.1. Most of the problems involved can be traced to growing demand for traditional fuels, coupled with diminishing supplies. They are aggravated, however, by the low efficiency of fuel-use devices and by the basic problems of poverty and inequality that limit the access of poorer members of the community to the fuel resources that still remain.

To solve the domestic energy problems of the developing countries, action on a broad front will be required. Part of the solution will be to expand biomass production and increase the efficiency of use of biomass fuels. The approaches that appear to have the greatest immediate potential in this context are as follows :

- Introduction of improved cooking stoves
- Better use of forest resources
- Rationalisation of charcoal-making activities
- Biogas

These are discussed individually below.

a) Introduction of Improved Cooking Stoves

The simplest and most obvious approach to reducing fuel supply problems in the short term is to improve the design and efficiency of cooking stoves. Using low cost materials and locally available skills cooking stoves can be built which are substantially more efficient than traditional stoves or open fires - when tested under laboratory conditions. If this increased efficiency could be translated into practice, under field conditions, it could lead to considerable fuel savings.

Designing improved stoves is complicated by a number of factors, however. For them to be accepted by users they must be designed so that they match local cooking habits and are convenient to use with the available fuels. They must also fulfil the subsidiary requirements of present stoves, such as room heating and keeping away insects.

Optimum designs will therefore vary from place to place and can only be developed through close interaction with the potential users - especially women, who are generally responsible for cooking.

It is important to note, however, that the introduction of 'improved' stove designs will not automatically reduce fuel consumption. If stoves are used incorrectly, for example, efficiency can be greatly reduced. Furthermore, even if stoves are used in the proper fashion people may prefer to use the same amount of fuel and just produce more heat.

b) Better Use of Forest Resources

In countries where substantial forest resources still remain, one of the simplest approaches to expanding fuelwood supplies is to increase the efficiency of utilisation of existing resources. Depending on the types of forest available and the location and nature of fuelwood demands, this could take a variety of forms such as :

- Increasing the utilisation of logging residues using portable or semi-portable charcoal making equipment
- Opening up new areas of forest for fuelwood extraction by relaxing legal constraints on forest use and improving access
- Adoption of improved forestry practices in which trees are replanted after cutting and measures are taken to prevent soil damage, forest fires, etc.

An essential feature of any scheme of this nature would be to prevent over-cutting so as to ensure the renewability of the resource.

c) Rationalisation of Charcoal-Making Activities

Charcoal is an important domestic fuel in many developing countries, particularly in urban areas, and is also used to a significant degree by industry. Charcoal production usually occurs in an informal and uncontrolled (often illegal) fashion, however, and the methods used are generally very inefficient - both in forestry terms and in energy terms.

So as to increase charcoal supplies, whilst at the same time ensuring the long-term renewability of wood resources, a number of approaches can be used :

- Regulation of charcoal making activities so as to reduce over-cutting
- Promotion of improved forestry practices (as above)
- Introduction of improved charcoal kilns to medium and large-scale charcoal-making operations (these can double the efficiency of the conversion process compared to traditional methods)
- Establishment of fast-growing tree plantations specifically for charcoal making. If unproductive forest areas are cleared for this purpose the income derived from the initial felling can help to offset the costs of plantation establishment
- Encouragement of charcoal-producing cooperatives so that small scale producers can combine their resources and purchase efficient kilns.

It should be noted that the impact of many of these measures will be limited in countries where charcoal production occurs mainly on a small-scale basis because the individuals responsible are often hard to reach, as well as being too poor to afford improved charcoal kilns (traditional earthen kilns and covered woodfires require virtually no financial cost). Regulation and control of charcoal-making will also be difficult in cases where illegal wood cutting is prevalent and where the vested interests of charcoal traders are strong.

d) Tree Planting for Fuelwood

In countries where fuel supplies are severely limited, planting fast-growing trees for energy purposes is the logical option for increasing biomass production. Tree growing can be performed on any scale, ranging from large plantations for supplying urban needs, to small village woodlots and patches of trees planted along roadsides and river banks and between fields. It can be organised commercially, through forestry organisations, or on a communal or private basis.

Given efficient cooking stoves and reasonable growth rates from wood plantations, domestic energy needs could, theoretically, be supplied from relatively small plantation areas. If annual domestic energy requirements were 7 GJ per capita, this would correspond to roughly 0.7 m^3 of wood. If a wood plantation yielded 15 m^3/ha/year, then approximately 1/20th of a hectare of plantation would be required per person to satisfy this need.

The feasibility of devoting such an amount of land to tree growing obviously depends very much on the population density in the particular area, and on the intensity of land use. In extreme cases, where population density is high and practically all the available land is already used for agriculture, this would pose major problems. In Bangladesh for example, where there are 6 to 7 inhabitants per hectare on average, 1/20th of a hectare per person would be equivalent to using 35% of all land for tree growing. Whilst not impossible (since technically there is significant scope for increasing crop productivity), this would require very major changes in agricultural practices.

More commonly, population density is not this high and reasonable amounts of unused or unproductive land exists which could, in principle, be converted to tree growing. There are a number of problems involved in using this land for tree growing, however, particularly during the early stages. These include :

- Low fertility and the resulting problems in establishing trees
- Ownership problems
- Competing uses for the land, such as for animal grazing
- The delay between planting and harvesting trees and the need to protect trees from animals, fire and other damage when they are young
- The lack of financial incentives if the wood is to be used only for fuel purposes.

Each of these factors will tend to increase the difficulties of establishing tree plantations, although there are a variety of ways by which the problems can be minimised. Where land is scarce, for example, Agroforestry techniques can allow food or fodder crops to be produced from the same piece of land by planting them in-between rows of trees. Likewise, the planting of multiple-use tree species that produce fruit, nuts, building materials or other marketable products can provide a financial incentive to tree growers. Where poor soil quality is a problem, leguminous trees can be used as a way of improving soil fertility.

Land ownership questions are more difficult to get round, however. Unless village woodlots are established or poorer members of the community are given rights to use common land, often it will only be the wealthier farmers who have the land or the resources that are needed to invest in tree growing.

e) Biogas

The production of biogas from animal dung and other organic residues has a number of theoretical advantages. It produces a fuel which is well suited for cooking, lighting and other purposes, and allows nutrients to be recycled to the soil. When used for disposal of human wastes it performs a valuable role in improving hygiene.

The technical feasibility of using biogas digesters for providing household energy is primarily determined by the availability of adequate quantities of digestible biomass material. Lack of water for mixing with the feed material can also be a problem in some locations, however, and the sensitivity of the process to low temperatures can rule out their use in areas where ambient temperatures drop below 10°C-15°C for significant periods of the year.

Assessing the relative costs of biogas digesters compared to other energy alternatives is a difficult process because it depends very much on the value assigned to the related benefits associated with biogas production, such as fertiliser production and health improvement. The design emphasis today is very much towards producing simple designs which can be constructed using local skills and materials, and this helps to keep costs down. Despite this effort, however, capital costs are still high compared with average rural incomes and for this reason the use of household scale biogas units is likely to be limited to the more wealthy families in a community. Furthermore, even if costs were reduced, poorer families would rarely have enough animals to justify building their own units.

To reduce capital costs and increase the scope of biogas technology so that poorer members of the community can contribute - and benefit - larger scale community biogas plants have some potential. These will require a substantial degree of village cooperation if organisational and distributional problems are to be avoided. They do allow the opportunity, however, of using excess gas for other useful purposes such as pumping drinking water, grain milling and electricity generation.

5.2 Energy for Rural Development

As was outlined in section 3.2, the availability of an affordable energy supply is an important component in the development of rural areas. The majority of the energy needs involved in rural development is of a relatively small-scale and decentralised nature; applications for which biomass energy is particularly well suited. In this section an outline is given of the main biomass options that appear technically appropriate for meeting these energy needs.

The most important theoretical limitation on the use of biomass energy for these purposes arises in the supply of adequate quantities of biomass materials - particularly since in many cases there are already shortages due to the lack of biomass fuels for meeting domestic energy needs. The first step in expanding the use of biomass energy to other activities, therefore, must be to develop new and renewable sources of biomass. The main options for achieving this are as follows :

1. Options for increasing the extent of use of present biomass resources

   - Increased use of logging and sawmill residues and improvement of forest management techniques
   - More extensive use of crop residues and food processing wastes
   - Harvesting of grasses, shrubs, aquatic weeds and other underutilised plant materials
   - Increased use of animal dung and human and household wastes.

2. Options for increasing the biomass supply by raising production

- Establishment of tree plantations
- Increasing production of agricultural crops and therefore crop residues using, for example, double cropping, irrigation, fertilizer, mechanisation, pesticides and interplanting with legumes
- Planting energy crops for the production of vegetable oils, ethanol or other fuels.

In almost all rural environments a significant increase in biomass productivity is theoretically feasible - although clearly this cannot occur without financial, agricultural and other inputs. Only in extreme cases where land is already highly productive or limitations exist due to factors such as fertility, water availability and climate, is the potential for increase biomass production only small.

Given that biomass can be supplied to rural communities at a reasonable cost and in adequate quantities, there are a variety of small scale utilisation options that are technically well suited to meeting the local energy needs involved in rural development. As already described, the precise nature of these needs varies and depends on the local agricultural system and development priorities. The main energy needs can be conveniently considered under three main categories, namely : stationary shaft power, mobile shaft power, and heat. The biomass utilisation options that appear most appropriate for meeting these end-use requirements are outlined below :

a) Stationary shaft power

Stationary shaft power is required for activities such as

irrigation, grinding of grain, small-scale industry and decentralised electricity generation. For reasons discussed in section 4.3.2 diesel and gasoline engines are probably best suited for small-scale applications while small steam engines come into their own for larger energy demands (25 kW and above).

To be used in internal combustion engines, biomass must first be converted to a compatible liquid or gaseous fuel. For most biomass forms, conversion of biomass to a gaseous fuel is much simpler than making liquid fuels and is, therefore, the logical option for stationary shaft power applications (where fuel storage and equipment size is less important). Suitable gaseous fuels can be made either by gasification of dry biomass materials (such as wood, charcoal and crop residues), using small-scale air gasifiers, or by anaerobic digestion of animal manure and other digestible organic matter to biogas. Diesel engines require only minor modification to burn these fuels (but employ small amounts of diesel fuel to provide ignition). Gasoline engines require major modifications, on the other hand, but can subsequently burn biogas and gasifier gases undiluted.

For larger scale applications, as mentioned, steam engines have advantages in that they can burn biomass directly. Wood or any other form of dry biomass would be suitable in such cases.

b) Mobile shaft power

Although the demands for mobile shaft power in most rural areas are still relatively modest, these demands will increase as rural development proceeds and transportation and mechanisation become more important. In view of the likely problems in supplying oil-based fuels there will be obvious advantages if the necessary energy can be produced locally, at a competitive cost.

The main options for supplying mobile shaft power in a rural setting are to use small charcoal gasifiers coupled to modified diesel engines or to produce liquid fuels on a small scale basis. Growing oil producing crops such as sunflowers, groundnuts and oil palm is probably the simplest approach to liquid fuel production. The oil from these crops can be extracted using presses and used directly as a diesel extender. With further treatment, these oils can be used undiluted as a diesel substitute. Small-scale production of ethanol also seems to be an eminently feasible option in cases where suitable land is available for growing sugar or starch crops. There is relatively little experience in either of these concepts as yet, however, since the majority of research work to date has been directed at large scale systems.

c) Heating Requirements

Heat is needed in rural environments for a variety of agricultural and industrial purposes. These include food drying and curing, tobacco curing, brick-making, brewing and metal forging. Heat can be supplied from a variety of biomass sources by direct combustion, wood being the most convenient fuel for most applications. For processes which need higher temperature heat, charcoal is an effective fuel.

5.3 <u>Energy for Petroleum Replacement</u>

The need to develop indigenous energy alternatives to replace imported oil was discussed in Section 3.3. Biomass energy offers a range of possible options for achieving this that could have important impacts in a number of developing countries.

As with the previous category of providing energy for rural development, the main technical limitation on using biomass energy techniques for petroleum replacement is that of biomass supply. To have a major impact in reducing petroleum requirements significant land areas are required and this inevitably restricts the potential of such schemes to those countries which have sizeable areas of unutilised or underutilised land.

To grow biomass on a large scale, to convert it to forms in which it can substitute petroleum, and to transport it to areas in which the demand exists, will require substantial investment of financial as well as other resources. It will also involve moderate and in some cases high levels of technical sophistication. While these factors will present obvious problems to any country with limited economic resources and technical capabilities, they are offset by the significant financial benefits that will accrue through replacing high priced petroleum with indigenously produced fuels. The economic constraints on system choice are, therefore, much less severe than for the previous two energy need categories in which low initial costs is a dominating criterion.

The available options for replacing petroleum with biomass energy are summarised below, having been introduced in more detail in Section 4. These options are divided into applications which can provide liquid fuels, energy for electricity generation and industrial heat.

5.3.1 Liquid Fuels

A wide variety of theoretical approaches exist for the production of liquid fuels from biomass. Of these the techniques which are available for implementation now or in the relatively near future, are listed as follows :

a) Production of ethanol by (yeast) fermentation of sugar crops, starch crops or surplus molasses

b) Production of methanol from wood, through gasification to produce synthesis gas followed by catalytic conversion to methanol

c) Production of gasoline from wood using gasification as above, coupled with Fischer-Tropsch conversion to gasoline. Alternatively, wood-derived methanol can be converted to gasoline using the 'Mobil' process.

d) Cultivation of oil producing crops such as oil palm, sunflower, coconut, sesame and groundnut.

The relative suitability of these various liquid fuel options under a particular set of conditions is determined by a wide variety of site-specific factors as well as policy decisions. Technical feasibility obviously relies on being able to grow biomass efficiently, which in turn is dependent on the type of land available, the climate, and the cultivation methods used. Theoretical liquid fuel production levels can be very high under optimum conditions. A comparison of optimistic production figures (on a GJ per hectare basis) for the main liquid fuel alternatives are shown in Table 5.1. These figures have limited relevance in practice, however, since in many cases the land that is available for growing energy crops is of

TABLE 5.1 Liquid Fuel Yields from Biomass under Optimum Conditions

Fuel	Feedstock	Feedstock Yield t/ha/yr	Conversion Yield % by weight	Fuel Yield t/ha/yr	Heat Content GJ/t	Fuel Yield GJ/ha/yr
ETHANOL[1]	Sugar cane	50	6	2.8	27	76
	Cassava	12	14	1.7	27	46
VEGETABLE[2] OILS	Oil Palm	25	20	5.0	40	200
	Groundnut	1.3	45	0.6	40	24
METHANOL[3]	Wood	20	30	6.0	20	120
SYNTHETIC[3] PETROLEUM (Fischer-Tropsch process)	Wood	20	14	2.7	40	108

Sources : Calculated from

1. World Bank (1980) "Ethanol Production in the Developing Countries".
2. J.W. Purseglove (1968) "Tropical Crops; Dicotyledons and Monocotyledon"
3. J.M. Radovich (1980) "Liquid fuels from Biomass - A Preliminary Process Comparison"; Proc. Conf. Bio-Energy '80, p.219.

relatively poor quality and actual crop yields will tend to be considerably lower. In addition, it may not always be economically or practically feasible to use the intensive cultivation techniques that are required to maximise biomass yields.

As well as the choice of biomass crop cultivation technique and conversion approach there is also a range of possible scale and organisational options for liquid fuel production. Some of the factors and trade-offs that are relevant in the selection and comparison of systems for liquid fuel production under particular circumstances are as follows :

- Economies of scale from building large conversion facilities compared to increased biomass transportation costs
- Labour intensive/capital intensive alternatives
- Requirement for skilled personnel
- Need for infrastructures
- Environmental impact
- Commercial/government ownership options

In view of the complexity of the system selection process, no overall judgement can be made as to which options are most suitable for developing countries, although some general observations concerning the technical sophistication and system versatility of the main alternatives are presented in Table 5.2. Furthermore, it is clear that many adaptations to the basic liquid fuel production approaches are possible - for example, modifying biomass production methods so that they are suitable for adoption by small farmers. As yet, many of these alternatives have not been fully explored; but they could have an important impact in making energy systems (which were often

TABLE 5.2 System Versatility of Liquid Fuel from Biomass Options

		PRODUCTION SYSTEM			CONVERSION SYSTEM		
Fuel	Feedstock	Agronomic[1] Requirements for deriving high yields	Range[2] of soil types tolerated	Drought Tolerance	Relative Capital Cost	Relative Technical Sophistication	Suitability for small-scale Applications
ETHANOL	Sugar cane	strict	narrow	low	moderate to high	moderately complex	Yes[4]
	Cassava	moderate	wide	high			
VEGETABLE OILS	Oil Palm	strict	wide	low	low	simple	Yes
	Sesame/castor groundnut etc.	strict	wide[3]	high			
METHANOL[5]	Wood	moderate	wide[3]	high	high	complex	No
SYNTHETIC[5] PETROLEUM (Fischer-Tropsch process)	Wood	moderate	wide[3]	high	very high	very complex	No

Notes : 1. Refers to fertilizer requirements and the need for irrigation, pest control, etc.
2. Refers to the ability of crops to grow on poor soils.
3. Different species will be suitable for different soil types.
4. Small-scale production of ethanol for fuel is technically feasible but not yet fully proven.
5. Technically feasible but as yet unproven on a large scale.

designed around the economic and social conditions prevalent in the US) more appropriate for use in developing countries.

5.3.2 Electricity Generation

The continually rising price of oil makes the generation of electricity from wood grown in fast-growing tree plantations an increasingly attractive alternative in countries where land is available for tree growing. Wood-powered electricity generation is generally best suited for extending the generating capacity of a country, as opposed to replacement of oil in presently existing oil-fired facilities (this is because of the transportation problems of supplying wood to existing facilities, which are commonly located in urban areas). Wood-powered plants can either be linked to national grids, or used to supply local needs in a small-scale decentralised fashion that would circumvent the significant transmission costs involved in providing electric power to a dispersed rural population.

The technologies involved in wood-powered electricity generation are well advanced - equipment choice depending mainly on the size of the facility. For small-scale units, diesel engines coupled to wood gasifiers are the most efficient option. For larger power requirements steam turbines or gas turbines linked to gasifiers are more appropriate.

The plantation size required to supply a wood-powered plant obviously depends on the plant capacity and the plantation productivity. As was the case for liquid fuel production, productivity and costs are very site-specific and a wide variety of scale and system options are possible. Feasibility studies have been performed in the Phillipines[69] and in Tamil Nadu State in India[70] into the possibility of generating electricity from plantation-grown wood. Plantation area was estimated at 110 hectares per MW of generating capacity in the Indian study (wood yield 55 ODT/ha/yr)*, and between 120 and 370 hectares per MW in the Phillipines (wood yield 16.5-50 ODT/ha/yr) - Where plans are already underway for building a number of plants based on this principle.

5.3.3 Industrial Heat

Most dry forms of biomass can be burnt very efficiently in well designed combustion equipment, making the replacement of oil in its industrial heating applications eminently feasible in technical terms. Given suitable land, biomass can also be grown at a cost well below the equivalent energy cost of oil - fast growing tree plantations being generally the most attractive option. Limitations arise, however, due to the problems of transporting biomass from the areas where it can be grown to the industrial centres where the energy demand exists.

Because of the transportation component, the most economic and energy-efficient method of using biomass for industrial heating is to locate industries near to energy plantations so that they can burn wood directly. This provides a strong incentive for decentralisation of new industrial facilities. Relocation of presently existing facilities, on the other hand, would obviously be limited by the economic and social upheaval that would be required.

Where haulage distances are large, conversion of wood to charcoal has a number of advantages. Its higher energy content (approximately double that of wood, on a weight basis) makes it cheaper to transport as well as more convenient and efficient to burn. Charcoal also has several other desirable features for specific industrial uses, such as steel and cement making. The primary disadvantage of using

---

* ODT = oven dry tonnes

charcoal is the significant energy losses that occur in the conversion process. These amount to as much as 50% of the original energy in the wood, even with efficient charcoal making equipment. Conversion to charcoal does greatly increase the range and extent to which wood energy can be used for industrial purposes, however, and as such can by justified given favourable local economics and provided that wood can be supplied in a renewable fashion.

The cost effectiveness of using wood and charcoal for industrial heat can be expected to improve as oil prices rise. The fact that these fuels are presently used in many countries is indicative that this is already a viable proposition in some cases, although the forestry and charcoal-making techniques currently used are often far from ideal. To maximise the potential of this approach, managed tree plantations will be required, as well as efficient charcoal conversion facilities and an appropriate transportation network. Such a supply system could either be integrated or separate from the systems that are necessary for the supply of wood and charcoal for domestic needs in urban areas.

Chapter 6

Implementation

The need to develop alternative energy strategies in developing countries is clear and in many cases it is also extremely urgent. Amongst the various energy alternatives, renewable as well as non-renewable, the modification and extension of the use of biomass energy has many advantages. As has been discussed, a wide variety of biomass options are available that appear technically feasible. Many also have the potential of yielding related benefits beyond their simple impact on energy supply; benefits such as environmental improvement and alleviation of the economic and physical hardships that are often characteristic features of present energy supply methods.

However, as has been stressed throughout this report, the gap between apparent technical and economic feasibility and successful implementation is often very large. In this chapter, the problems of achieving successful and effective implementation of biomass energy systems are discussed. As well as the development and transfer of technology, which are vital in themselves, devising effective implementation strategies also requires a full understanding of many micro-level economic and socio-political factors. These factors - attitudes, perceptions, priorities, power structures, and other features within recipient groups - play a crucial part in determining the rate and extent to which new technologies are accepted and the impacts that they subsequently have, both good and bad.

In the initial part of this chapter, observations are made relating to implementation efforts that have been made to date in introducing biomass energy systems. The main success stories are highlighted but the frequent lack of success of many projects is also noted and some general lessons are drawn regarding some of the common reasons for failures. Subsequently, a brief discussion is presented of some of the inherent features within developing countries, at the micro-level as well as at higher levels, that influence the acceptance and rate of adoption of new energy systems and present constraints on the immediate impact of biomass techniques. Next, some comments are made concerning underlying problems that hinder the spread of new biomass schemes and possible ways around them. Finally, a tentative list is drawn up summarising elements which seem essential to the design of effective implementation programmes.

6.1 Implementation Experience to Date

Many of the biomass technologies that have been discussed are not new and in a number of developing countries serious attempts have already been made to implement biomass energy schemes. Up to now, the majority of these efforts has been directed at solving household energy problems and has used techniques such as improved cooking stoves, biogas and community woodlots; the main exceptions being the measures that have been taken in a few countries to introduce improved charcoal kilns to forest industries and to promote the production and use of ethanol for fuel.

Although there are some notable exceptions, in general the success of implementation efforts to date has been decidedly modest. Equipment has been built but never used, or it has been introduced but subsequently abandoned. Where new ideas have been accepted, they have rarely spread as fast as was expected. In some cases, new technologies have had actively detrimental effects, bringing benefits only to the wealthy and making the poor worse off.

Whilst partial and slow progress has been reported in a number of countries, only in a handful of cases have new biomass concepts 'taken off' to the extent that they have had a major impact on the energy supply situation. The Chinese experience with household biogas units and to a lesser extent reafforestation (for environmental protection as well as fuel wood), and the South Korean community woodlot programme are the most striking 'successes' in the household energy sphere. For larger scale systems, the Brazilian ethanol programme stands out as having made an impressive start in reducing petroleum imports. Even these have had problems and side-effects, however, and can be argued as having been only possible due to special circumstances - political or social systems for example, or in the case of the Brazilian ethanol programme, exceptional resource endowment coupled with particularly severe balance of payments problems and a slump in world sugar prices.

Although a considerable amount of effort has been devoted to developing and implementing biomass energy systems, very few detailed studies have been performed into the factors which affect the rate of adoption of biomass systems or their subsequent impact. What evidence that has emerged, however, has shown that the diffusion of new energy technologies is a far from simple process and that it is likely to meet with many of the same problems that have hampered other rural development programmes - such as those aimed at introducing high yield crop varieties, irrigation systems, public health measures and population control.

The successes and failures of specific biomass energy implementation efforts are not commented on in detail here. It is relevant, however, to review some of the common problems that have been encountered since a number of important general conclusions can be drawn from these regarding what is required for a new energy system to be effectively implemented. These general problems are dealt with below under a number of heading - although clearly they are often interrelated in many ways.

a) Lack of resources and inequalities

Poverty, and the lack of resources for investment in energy systems, have been an obvious impediment to their diffusion in many cases. Inequalities in the ownership of resources and in general political and economic power within communities, have also meant that frequently it is only the rich that are in the position to adopt and benefit from energy innovations.

This problem has arisen, for example, in the introduction of biogas plants. To own a household-size biogas digester a family needs both the financial resources for constructing it and sufficient animals to feed it, once built. Studies in India, Thailand and Korea, have shown that lack of resources has been one of the major reasons for non-adoption of biogas plants.[78] A survey in Gujurat State, India,[52] also indicated that the majority of families owning biogas plants have much larger amounts of land and cattle than the local average. A similar pattern of uneven and limited impact was observed in a case study of a tree planting scheme in India[9] - lack of land, labour, time and water being the primary constraints preventing the poorer members of the community from participating.

It should also be noted that where inequalities within a community mean that only the wealthy can adopt new energy systems, the impact of these systems can sometimes be actively detrimental to the poor. Monetisation of dung and firewood, for example, which may result following the introduction of biogas or tree planting schemes, can create very serious problems for the poor - who previously collected these fuels free but are excluded from participating in the new scheme due to their poverty or lack of political power. Likewise, the privatisation of land as part of a controlled forestry programme can deprive the poorer sectors of the community of land for grazing and fuel collection unless measures are taken to supply alternative sources (at an acceptable price).

b) Design and reliability problems

In a number of cases it has been found that energy systems which have performed well under test conditions have proved unworkable or unreliable when introduced into the field. Examples of this include metal gas-holders on biogas digesters that have been subject to rapid corrosion[78] and the short lifetime of portable metal charcoal kilns[40] - again due to corrosion problems.

Inappropriate design has also been cited as a common problem in the adoption of supposedly 'improved' cooking stoves. In Ghana, for example, new stoves were abandoned because they were less convenient than traditional models and because they were not suitable for use with existing pots.[43] In Guatemala, early efforts to diffuse 'Lorena' stoves ran into problems because, amongst other things, they did not provide adequate room heating and were less versatile than the open fires used previously.[71] Some householders, in fact, got round these disadvantages by removing the firebox door. Whilst this increased the amount of space heating given off and permitted direct cooking inside the firebox, it totally undermined the design principle behind the stove which was to minimise fuel consumption.

c) Lack of interest and cooperation from local people

The success of many programmes has been severely limited by the lack of interest and participation by recipients. A number of superficial reasons have often been quoted to explain this phenomenon such as apathy, lack of education, slowness to change, out-moded traditions and apparently illogical behaviour by local populations.

Whilst these explanations may sometimes be accurate when viewed by the outside observer, many of them are merely reflections of underlying factors which are often determined by more logical considerations. For example, factors such as the avoidance of risk, the inconvenience of new designs, the relative unimportance of energy concerns as compared to other more pressing problems (such as food), and conflicts of interests between members of the community, can mean that - when seen from the point of view of the user - a decision not to adopt a new energy practice may be entirely logical.

Lack of interest and non-cooperation have been reported, for example, as the major cause of failure of a number of West African woodlots projects.[43,75] Even when community participation has been gained in some cases in planting trees, trees have been allowed to die due to lack of care. In others, saplings were willfully destroyed by the local population. The reasons for these actions are often complex (involving animal grazing rights, lack of involvement of women, lack of appreciation of fuelwood problems, and other factors), but they indicate how well-intentioned programmes can go completely wrong due to a failure to understand and accommodate the needs and attitudes of local people.

d) <u>Ineffective extension services</u>

A substantial number of the problems involved in achieving successful implementation of biomass energy (and other) schemes can be put down to the ineffectiveness of government extension services. Lack of manpower and resources are an obvious constraint in many countries but often there is a range of more general problems which prevent the services that countries do have from being more effective. These problems include, for example, a low level of commitment from extension workers, unwillingness to consider and respond to the needs of users (especially women), poor coordination from above and in some cases, corruption. Each of these factors tends to create a situation where local inhabitants are sceptical of government programmes and government personnel and are unwilling to participate fully unless it is very obvious that the benefits will go to them.

The ineffectiveness of extension services can frequently be traced to the 'top-down' approach that has commonly been taken in the past - in which user needs (or national needs) are assessed from outside and then efforts are made to design and diffuse 'appropriate' energy solutions. Experience has shown that this kind of approach often results in a failure to comprehend the problems of local inhabitants, and a failure to appreciate their requirements for 'appropriate' solutions and elicit their trust and support in carrying out programmes.

6.2 General Features of Developing Countries relevant to Implementation of Biomass Energy Schemes

There are many lessons that can be learnt from the successes and failures of energy implementation efforts in the past. Certainly, it is clear that energy supply patterns and their relationship to other aspects of the social and economic system are often much more complex than might be predicted from a superficial technical or economic analysis. Likewise, the organisational and support measures that are necessary to sustain an implementation programme go far beyond the simple supply of technologies or hardware.

To assess the true feasibility of biomass energy in the short term, therefore, and to identify the key requirements for effective implementation in the future, it is vital to understand the realities of the social, economic and political system in developing countries. Whilst it goes without saying that these 'realities' vary tremendously between countries and that they are constantly changing under the influence of modernisation, population growth and other factors, coming to grips with them is probably the single most important precondition for effective progress in solving the energy problems of the developing countries.

In this context and at the risk of over-generalisation and over-simplification, some of the important features of the present systems within developing countries are outlined in the following sections. A division is made between 'micro-level social and economic features' and what is termed 'infrastructural and macro-level considerations'.

6.2.1 Micro-level Social and Economic Features

To assist in the understanding of local energy problems and to provide insight into how new energy schemes will be received, it is important for planners to have some impression of the criteria on which local people base their decisions - and the contraints placed on these decisions by social and economic factors. Clearly, traditions and behaviour patterns differ tremendously both between and within countries. However, so as to provide a basis for further discussion, an attempt has been made below to highlight some of the attitudes and social characteristics that are prevalent within developing countries (particularly the poorer sectors) and are important in determining the acceptance of new energy schemes. This crude, 'thumb-nail sketch' is divided into a number of categories, as follows :

- perception of energy problems
- importance of traditions
- costs and benefits
- power structures and inequality

The examples that are taken to illustrate the points that are made are mainly drawn from experience relating to implementation of household energy schemes. This is not intended to negate the importance of micro-level considerations in determining the success of larger scale biomass energy programmes (they are likely to be equally crucial) but reflects the present lack of experience in implementing these schemes and the lack of knowledge as to what the problems might be.

a) Perception of energy problems

For the energy analyst, from his external vantage point and with his scientific knowledge of the effects of deforestation, nutrient depletion and population growth, the energy problems of the poorer sectors in the developing countries may seem very clear. This perception is not always shared by the local populations affected by these problems, however, and this difference in viewpoint can often explain their apparent lack of interest in new energy schemes. For the family that can still find fuel from one source or another, without a huge expenditure of time or money, there may be a number of far more pressing concerns to think about - such as providing tomorrow's meal - rather than worrying about the long term impact of their fuel collection habits or the thermal efficiency of their stoves. Thus, even in situations where the environmentalist can see impending ecological crisis, it is not uncommon to find wasteful use of fuel resources such as deliberate starting of brush fires and allowing cooking fires to die down naturally rather than dousing them to conserve fuel. In a similar way, assuming a farmer has access to limited amounts of capital he may choose to invest it in, say, keeping another pig - rather than building a biogas digester - because an extra pig may bring him short-term profits whilst a digester can only yield benefits in the longer term.

b) The importance of traditions

Traditions are often one of the major impediments to change in any society and in the rural areas of the developing countries traditions are particularly strong. Whilst some traditions are clearly anachronistic, others are reflections of an underlying rationality that has evolved over many years to cope with the problems of survival in an often harsh environment. The fact that they have persisted is, in a sense, proof that they have been (on the whole) successful.

Traditions are often deeply ingrained in the way of life of rural populations, playing an important role in maintaining cultural integrity and community cohesion. For this reason, changing traditional practices can often have a variety of hard-to-predict side effects, not all of which are beneficial. Changing the traditional patterns of fuel collection and used is probably a necessity, however, if rural energy problems are to be solved. In view of the deeply-held nature of these traditional practices, however, this change will inevitably be slow and new schemes must be demonstrably attractive if they stand a chance of being accepted.

In addition to traditions governing current energy practices, taboos can also grow up regarding the adoption of new schemes. The classic example of this is the resistance that has been met in many parts of the world to the use of animal dung and human excreta for biogas generation. For populations who have 'traditionally' collected and used animal dung for cooking, such as in India, use of dung for biogas production is likely to be less of psychological 'jump'. In others, however, where (as in the West) human and animal faeces are regarded as 'dirty', these problems should not be under-estimated and much time and persuasion will probably be required to overcome these inhibitions.

c) Costs and Benefits

To appreciate the reasons why energy schemes have often been rejected by local populations it is important to understand the means by which the individual within the community arrives at decisions. Although the exact reasons may not be obvious to the external observer, people base most of their decisions on what they perceive as the costs and benefits of a particular action - as compared to other possible actions. These are the costs and benefits viewed through their eyes, however, a view which can often be very different from those perceived by the outside observer. They can also vary significantly between members of the same community because of the different degree of access that individuals have to necessary resources and potential benefits.

The cost of adopting a new energy practice can be viewed in terms of the direct financial and labour inputs that will be required to bring it about, and the relative significance of this cost will obviously vary depending on the ability of the individual to provide these inputs. As well as direct costs, however, there are indirect costs which relate to the benefits that will have to be foregone by taking that particular action - as opposed to doing nothing, or adopting another course. In situations where resources are limited these indirect costs are liable to be as important as the direct ones, although they are difficult to quantify in financial terms. As an illustration of this point, if one considers the costs to a farmer of converting some rough grazing land to a tree plantation the direct costs will include seedling costs and labour requirements. Indirect costs, on the other hand, will be related to the problems of replacing the animal feed, building materials and firewood that this land had previously supplied, as well as the benefits he will forego by planting fuelwood trees on this land as opposed to, say, fruit trees. (Such an overall cost of a particular action is referred to as the 'opportunity cost' in formal cost-benefit analysis.*)

The benefits that accrue to an individual from adopting a new energy practice can also be of a direct or indirect nature and can occur over various timescales. If a new energy scheme results in the

---

* For a review of cost-benefit analysis of biogas schemes see Barnett (1978)[7]. A financial appraisal of several renewable energy projects is presented by French (1979)[33].

production of a saleable product (such as a commercial fuelwood plantation), or the reduction in consumption of a purchased fuel (such as an improved stove to an urban dweller), benefits will be of a direct financial nature. This will improve the overall desirability of the scheme considerably, since it provides a means of paying back the original investment and gives a direct incentive to the user. Such schemes will also be more susceptible to diffusion by normal market mechanisms.

More problematical, however, are those schemes which yield no direct or immediate financial gains and only more diffuse benefits such as increased conservation of fuel, convenience, environmental improvement, gradual increases in soil fertility or better public health. This situation applies to many domestic energy schemes directed at rural populations who presently gather fuel as a 'free good'. While fuel conservation would obviously be a great asset if present fuel collection practices are very laborious, it would be a much less compelling argument if the burden of gathering fuel is only moderate in comparison to the many other daily tasks that must be performed (as is more commonly the case).

The importance of convenience in determining the acceptance of new energy practices appears to be particularly great in relation to schemes which effect the nature of the cooking process. For example, in India it has been reported that one of the primary motives for purchasing biogas plants has been that the clean-burning gas was preferred for cooking to the smokey flame given off burning cow dung (which blackens the kitchen, utensils and clothing, and can produce serious eye infections).[78] On the other side of the coin, some attempts to introduce improved cooking stoves have run into problems because of their decreased convenience compared to traditional models (involving, for example, cutting wood into smaller pieces, increased care of the fire to ensure maximum efficiency, or the purchasing of new cooking utensils). Designing stoves which combine fuel conservation with greater convenience is now recognised as an important feature in effective stove implementation since, as it has been put in the paper by Celeski et al,[19] "people in developing countries are likely to be no more willing than those in developed areas to change habits they enjoy or find convenient in order to conserve energy or use it more efficiently". Another feature that has been realised in relation to the introduction of new cooking methods is the important role that women play in determining the acceptance of these ideas. Women are integrally involved in the fuel gathering and cooking processes and it is frequently the costs and benefits through their eyes that determines the rate of adoption of new methods; although it is also dependent on the degree to which women take part in decision making over financial matters.

One of the common problems relating to renewable energy schemes is that long term and indirect benefits are often either not perceived by the individual, or are overshadowed by the short term costs. To the government, on the other hand, these are often the primary motive for promoting a new energy scheme. A further problem is that, where indirect benefits are significant, they don't always necessarily accrue to the individual who pays the cost of adopting a new energy practice. Thus, for the landless peasant, the benefits derived from more efficient recycling of soil nutrients (through use of a biogas digester, for example, or from switching away from crop residues as a fuel) are likely to go primarily to the landowner in the form of improved crop yields. On the larger scale, many of the benefits from adopting more ecologically sound forestry practices in upland regions go to the populations living downstream, in the lowland areas, in the form of reduced flooding and siltation. In both cases, unless the direct benefits to the individual who pays the cost of adopting a new energy practice are also significant, overall motivation is likely to be very much reduced.

d) Village Power Structures and Inequality

From the experience that has been gained to date in implementing rural energy schemes (as well as other innovations), it is becoming increasingly clear that factors relating to political power structures and inequalities at the local level are extremely important in determining who can adopt new energy schemes and who will benefit from them.

The nature of power structures and the degree of inequality, varies tremendously between different parts of the world. In Asia and Latin America inequalities are particularly severe, being reflected primarily in land ownership patterns. For example, in Latin America as a whole, 70% of cultivated land is owned by less than 2% of all landowners.[81] In Bangladesh, 6% of landowners own about 54% of the land.[41] At the bottom end of the equality scale are the landless, who in many countries represent more than a quarter of the rural labour force and the near-landless who own only small plots of land. These groups scrape a highly tenuous existence through share-cropping or hiring out their labour for very meagre wages, and often exist in severe debt, unable to receive credit except at exhorbitant interest rates.

Under these conditions of extreme inequality, the scope for poorer members of the community to adopt and benefit from any new energy scheme is minimal if that scheme requires any financial investment at all. Control over the use of land and other resources and access to credit, are all in the hands of the wealthy. It is therefore not surprising that it is these groups that have generally been in the position to adopt new energy technologies in the past, thereby deriving the benefits for themselves and strengthening their relative position - in a self-reinforcing fashion. Even when subsidies and government assistance are made available, with the aim of helping the poor, it is often the more powerful members of the community that benefit most since they can control the means of distribution of aid and have the additional resources necessary to capitalise on partial grants or subsidies.

The situation with regard to land ownership is notably different in many parts of Africa. Except in former colonial settlements where plantations were established to grow coffee, sugar and other cash crops, land ownership is generally of a less formal nature than is found in other parts of the world. Population density is lower and the majority of peasant agriculture occurs on a subsistence basis, commonly involving practices of shifting cultivation. Land is allocated to individuals by community chiefs in accordance to needs and is not formally owned or passed down between generations. Land ownership, therefore, is generally not such an important determinant of wealth and power - these relying more on features such as the number of animals owned. Communal forces are probably stronger than in most Asian or Latin American countries and although absolute levels of poverty are some of the worst in the world, inequalities within rural populations are less marked than in other countries. Under these conditions, therefore, one would expect that inequality (at the local level) will be less of a constraint to the introduction of new energy schemes than poverty itself, or other factors such as traditions and the lack of technical and financial infrastructures.

6.2.2 Infrastructural and Macro-Level Considerations

As well as factors relating to the eagerness and ability of individuals to accept new energy systems, the rate of adoption of these systems is also constrained by the capacity of the nation to undertake and sustain a major implementation effort. Once again this varies tremendously between countries but in many cases is a crucial factor in determining the immediate prospects for implementing biomass energy concepts. It also has implications with respect to which biomass systems are most suitable under local conditions, due to the varying requirments that different systems have for infrastructures, resources and other inputs.

The main features of developing countries relevant to their capacity to implement effectively biomass energy schemes are outlined briefly below under the following headings : skills and infrastructures; extension services; and planning and research capabilities.

a) Skills and Infrastructures

To introduce new energy schemes to developing countries certain skills and infrastructures are required, the nature of which depends very much on the particular technology being introduced. These include :

- skills in building and using new equipment
- knowledge of improved agricultural techniques
- local manufacturing and maintenance capabilities
- credit facilities
- transportation networks

The extent to which these capabilities vary is highlighted when one compares countries like Brazil and India, which are relatively advanced in technical terms, with some of the less technically developed nations. In Brazil and India the facilities that exist in small towns for manufacturing and repairing even quite complicated equipment are very good, at least in the more affluent regions, and banking and government sponsored financing are already established (although they may not always work ideally). In contrast, the levels of skills and other infrastructures existing in some of the poorest nations are very low and this effectively rules out the adoption of sophisticated energy technologies in the short term.

b) Extension Services

As has been mentioned, one of the key factors in ensuring the success of any implementation scheme is the provision of effective extension services. These services form the interface between government planning and research capabilities and the actual energy user and are therefore required for each of the following, crucial roles :

- Perceiving the energy problems of users and understanding the constraints that they face in relation to the adoption of new energy schemes
- Determining what energy demands are (as opposed to energy needs), and ascertaining what system features are required for energy innovations to be attractive to users
- Stimulating local participation in the development and implementation of energy schemes
- Demonstrating new techniques, training local people and providing advice and back-up
- Monitoring of schemes to assess their progress and helping in recommending possible improvements.

Many of these roles can only be performed effectively through close interaction with the people involved. For this reason, a 'bottom-up' approach to the diffusion of new energy systems is generally much more likely to succeed than a 'top-down' approach - in which the diffusion process is essentially a one-way transfer of 'appropriate' energy solutions which have been developed in isolation from the actual users. A 'bottom-up' approach is not easy to achieve in practice, however. It involves organising and assisting people to solve their own energy needs, developing locally acceptable responses (technical and organisational) and carrying out mutually agreed programmes.

Cooperation is obviously vital, but for this to be forthcoming people must view energy schemes as being for their benefit. Developing confidence in government extension services is a crucial part in this and a combination of persuasion, demonstration and suitable incentives may often be required to bring about the necessary transition in people's attitudes. Employing women extension officers can be useful in this context as it can help to encourage the participation of women in developing and carrying out energy programmes. Organising energy programmes through schools, temples and other well-respected bodies can also help to gain the trust of local communities.

Depending on the particular programme, extension services will require personnel trained in a range of specific technical subjects as well as social workers and teachers. In many cases the levels of training required for extension workers is fairly modest. What is often more important is the degree of dedication of these workers, effective leadership and organisation from above, and willingness to work with - and listen to - local people.

Unfortunately, very few of the developing countries can boast of having extension services that are either of the size or the quality necessary to undertake any major biomass implementation effort in the near future. Setting up efficient extension services is, therefore, one of the first steps that must be taken in promoting biomass energy and this is one area where much could be learnt from the Chinese experience. in rural development.[5]

c) Planning and Research Capabilities and Legal Constraints

As well as infrastructures and extension services, strong and coordinated national leadership is required for effective implementation of new energy schemes. This is necessary for functions such as :

- overall energy planning
- coordination of extension services
- organisation of basic and applied energy research.

Once again, relatively few developing countries have sufficiently strong or well organised central institutions to perform these roles effectively. More often, the relevant government ministries suffer from a lack of personnel and the degree of cooperation between them is frequently very poor. Many of the important fields, such as forestry and agriculture, also have a low prestige associated with them which makes it difficult to attract high calibre personnel to these jobs.[1] These situations must be remedied if countries are to make full use of the personnel and resources at their disposal.

On the legal side, official regulations regarding the use of land often present constraints on the development of certain types of biomass energy schemes. For example, access to government forests is often restricted in an effort to protect the resource from indiscriminate logging and deforestation. While these efforts are rarely totally successful the laws themselves will have to be modified if the opposite tack is to be taken - that of maximising wood production through controlled forestry.

6.3 Implementation - A Positive Approach

6.3.1 Underlying Problems and Possible Solutions

The previous sections have highlighted some of the underlying problems that hinder the spread of biomass energy schemes - factors such as poverty, inequalities, traditional values and the lack of effective implementation machinery. A number of basic contradictions also exist pertaining to changes in energy use practices - contradictions between food and fuel production, between short-term imperatives and long-term costs, between the needs of the individual compared to those of the community as a whole, and between the interests of the rural population as opposed to urban dwellers.

Many of these underlying problems and contradictions are inherent in the cultural, social and political systems that exist in the developing countries. Traditions, in particular, are deeply ingrained (and important) features of societies and as such will inevitably be slow to change. Likewise, poverty will not disappear overnight, particularly in countries where it is continually reinforced by the mechanisms of inequality that maintain present class structures. To alter this situation major socio-political changes will be required and these involve a whole range of issues that are beyond the scope of this paper (although they cannot be ignored).

Whilst these basic problems are of fundamental importance (and they should be recognised from the outset as such), they should serve to focus the attention of the policy makers and scientists on what can be done in practice to improve the effectiveness of implementation methods. In this context, a number of approaches can be taken to try and make biomass energy more attractive to the individual and to increase the availability of necessary financial resources. These are discussed below.

Technical Solutions

Although technical 'fixes' alone are unlikely to solve any of the more fundamental problems involved in introducing new energy systems, they certainly have an important role to play. Through a variety of technical measures it is often possible to adapt biomass technologies to be more 'appropriate' for a particular set of conditions and therefore increase their attractiveness to users. For example, in the design of hardware short-cuts can usually be taken to simplify the methods of manufacture and operation and to reduce the cost by using locally available materials and skills. This is currently one of the main thrusts of research and development into stove and biogas digester design. Design modifications can also improve the convenience of operation of equipment. Together, low cost and convenience are often the dominant factors in determining the acceptance of new equipment and as such can override the importance of system efficiency as the primary design criterion.

For biomass production systems, the food/fuel/fodder conflict is one of the major limitations, as is the long delay between establishing tree plantations and harvesting them for fuel. These can be ameliorated in a number of ways such as by interplanting food and fuel crops (Agroforestry), planting catch crops for energy purposes during fallow periods, using only very poor and underutilised land for growing energy crops or employing presently wasted biomass materials.

It should be noted that the scope for adapting biomass energy concepts to improve their suitability under local conditions applies to medium and large scale systems as well as the smaller scale options. This is one area which is sorely in need of further research, since many of the larger scale biomass energy systems have been developed for use under the conditions that exist in the industrialised nations, which are radically different from those in the developing countries.

Economic Measures

To increase the financial attractiveness of biomass energy systems and to bring them within the budgetary range of the poorer sectors of the community a number of economic strategies can be used. These include direct grants and subsidies, setting up of low interest revolving loan funds and modification of collateral rules so as to allow borrowers greater flexibility in what they can offer as a guarantee against loans (at present, land is frequently required for collateral - this discriminates against the poorer members of the community and often speeds up the slide into landlessness).

It must be recognised, however, that the impact of many of these strategies will be constrained by the power structures that already exist within societies. It will often be very difficult to design systems which are not susceptible to distortion by power groups and which really do allow assistance to reach the poorest members of the community. Devising ways of achieving this is likely to be a crucial part of arriving at real solutions to the energy problems of the poor.

Community Energy Schemes

The use of communal energy sustems is an idea which has many theoretical attractions.* By pooling land, skills and capital, communities can greatly increase the range of energy options open to them and they can make much more efficient use of the resources that they have. Community schemes can also permit poorer members of the community to participate and benefit from systems and technologies that they would be unable to afford otherwise.

Community biogas plants and village woodlots are two examples which illustrate the potential for communal energy schemes, as well as the problems involved in making them work. Clearly, a commonality of interests is vital if the approach is to be successful and an agreed and equitable system must be established for the sharing of costs and benefits. These are likely to be difficult to achieve in highly fragmented societies in which traditions of community cooperation are weak and where conflicts of interest dominate.

6.3.2 Requirements for Effective Implementation

A wide variety of implementation approaches are possible for biomass energy systems, involving various levels of government intervention and sponsorship and differing degrees of reliance on market forces and entrepreneurial activity. Clearly, no prescription can be written to guarantee effective implementation since the environment for implementation varies tremendously between countries - and what is effective for one particular energy system, in one particular location, may be totally inappropriate elsewhere.

---

* For a more detailed discussion of the role of public participation in rural development in general, readers are referred to the article by Cohen and Uphoff.[21]

Accepting this basic proviso, however, it is possible to delineate some very general preconditions for effective and widespread adoption of new biomass energy systems. These can be stated as follows :

- Energy systems should be appropriate for local conditions and attractive to potential users
- Systems should be affordable to a wide sector of the population
- Systems should be available along with the necessary infrastructures and programme back-up.

Based on the limited implementation experience to date it is possible to distill some essential elements which appear to be necessary (or at least helpful) in ensuring that these preconditions are met. These are summarised below under the heading of System Choice, Financing and Incentives, and Extension Services.

It should be noted that the majority of the points made refer primarily to small and medium-scale biomass energy systems; systems for meeting domestic and village level needs. Very little experience exists in factors relating to the implementation of large-scale biomass schemes (petroleum replacements technologies, for example) and these will not be dealt with in detail in this review. It is important to be aware, however, that a wide choice of system alternatives are available involving various scale options and degrees of private/government control. Social and economic impacts of large-scale systems will inevitably be great and choosing the most appropriate system option and implementation approach will be a vital part of ensuring that the maximum benefits are realised - in social and developmental terms, as well as environmental and economic ones.

System Choice

1. A thorough understanding of present energy use patterns and the various constraints on the expanded use and development of biomass resources.

2. Involvement of local groups during all stages of planning, selection and testing of new energy systems (particularly women's groups when it is women that are primarily affected by new techniques).

3. An understanding of energy (and other) problems as seen through the eyes of local inhabitants, so as to derive an insight into the relative costs and benefits to them of adopting new energy systems.

4. An appreciation of the power structure within communities and the likely limitations that these will place on the adoption of new systems.

5. Incorporation of benefits other than just energy efficiency into designs. This includes factors such as convenience, simplicity, and compatability with present energy use practices.

6. Minimization of financial costs and design of equipment so that it requires the least amount of external inputs in building, operating and maintaining the system.

7. Design of systems so as to yield direct financial benefits wherever possible.

Financing and Incentives

8. Provision of financial assistance in cases where spread of technologies is limited by lack of financial resources and credit facilities.

9. Targetting of subsidies and grants to poorest members of the community.

10. Provision of additional incentives (financial or otherwise) to promote adoption of energy schemes which may yield only marginal benefits to the local inhabitants but which have important wider-scale advantages.

Extension Services

11. Extension services should endeavour to set up good relations with local groups to win their trust and support.

12. Persuasion of local people as to the benefits of new technologies - using a combination of demonstrations, educational efforts and propaganda; where appropriate.

13. Links should be set up to encourage feedback from users and extension officers to higher level research and planning groups, to allow techniques and implementation approaches to be continually improved so as to match the needs of the local population better.

14. Provision of incentives to extension officers (financial or otherwise) to enhance their prestige and encourage hard work and dedication.

15. Biomass energy extension efforts should be tied in and coordinated with other development programmes (agricultural, forestry, industrial, public health, educational, etc.), at every level.

16. Setting up of training schemes to provide adequate number of appropriately trained personnel (the levels of training need not always be high but large number of officers will be required to mount a major implementation effort).

17. Effective leadership and coordination of extension services.

18. A major commitment from top-level government bodies to promote energy projects which is both visible and backed up through positive action.

Chapter 7
General Policy Considerations

In this final chapter some general policy considerations regarding the development of biomass energy are presented. First of all, the important linkages that exist between choices of biomass systems and overall energy and development policy are discussed. Subsequently, comments are made concerning some major areas where R and D into biomass energy methods is needed. This is followed by observations regarding the problems of coordination of biomass energy development and the opportunities that exist for constructive collaboration between concerned groups at various levels. Finally some general comments are made with respect to the potential for biomass energy in the developing countries.

7.1 Biomass Energy and Development Policy

Although this study has concentrated on biomass systems it must be remembered that biomass energy is only one of a number of energy alternatives for the developing countries and that energy policy is only one facet of overall development planning. In reality, there are many linkages between these policy elements and it is important that an integrated approach is taken so as to coordinate biomass energy development with broader scale policy considerations.

One of the most important aspects of planning energy strategies is recognising the broader implications of energy decisions. Different energy alternatives have very different characteristics; they require very different inputs and will have very different effects. In very few cases will a simple finanical appraisal of energy alternatives accurately reflect their full costs and benefits. Generally, a much more detailed analysis will be required, taking into account issues such as :

- Suitability of energy systems for supplying the energy requirements of the country - given the geographical distribution of demand, the nature and magnitude of demand, and changes over time
- Capital costs of energy systems, infrastructural requirements and the demands they place on scarce resources
- Foreign exchange requirement and the need for foreign finance, technologies and know-how
- Implications with regard to helping to achieve national self-reliance, rural development, economic growth or other development goals
- Suitability for meeting the energy needs of the poor
- Environmental impact

On the basis of these criteria, biomass energy systems clearly have very different developmental implications compared to most other energy options. The suitability of some biomass systems, for example, for low-cost small scale decentralised applications make them potentially a powerful tool in stimulating rural development. Not only can biomass provide energy for local uses in industry and agriculture, 'energy farm' systems can also be developed whereby production of biomass energy and its sale to urban areas provides a source of employment and income to rural dwellers.

It should be stressed, however, that biomass energy systems - in common with all other development strategies - do not in themselves guarantee equitable rural development. Unless efforts are made to channel the benefits from the system to the poorer sectors of the community and prevent profiteering by dominant groups, the situation could easily develop whereby the resources of rural areas are exploited purely for the benefit of affluent urban populations and to the detriment of those living in rural areas.

On the environmental front, biomass systems can have advantages in countries where the curbing of deforestation is a major development priority. It may be possible, for example, to reduce the costs of reforestation if tree planting is coupled with the establishment of plantations for energy uses since it can allow a direct financial payback to be derived from reforestation. Perhaps more importantly, such an approach to reforestation can also offer an alternative livelihood for the local inhabitants who have been responsible for forest destruction.

Overall, biomass systems form a very heterogeneous group. The simplest systems, such as improved woodstoves, have very few similarities with some of the larger scale options - methanol from wood systems, for example - which are highly sophisticated and require high levels of technological and financial inputs. Systems must therefore be assessed on an individual basis since their suitability under local conditions and their implications with regard to national development priorities will be very different.

Whatever systems are chosen, however, it is important that the various elements of national energy policy are coordinated so as to be mutually complementary. Pricing policies for electricity and fossil fuels, for example, will have a major effect in determining the relative economics of biomass options and their rate of acceptance. Strategies on pricing and/or subsidies must, therefore, be used in an integrated fashion if they are to have their desired impact. Likewise, the regional and national endowment of energy resources (fossil fuels, hydro, geothermal, etc.) must obviously be taken into account in developing biomass energy policies so that the optimum matching of energy sources to energy needs can be made.

7.2 Research and Development (R and D) Needs

At present, the vast majority of biomass energy R and D is concentrated in the developed countries and is focussed on solving the energy problems of those countries. Whilst this undoubtedly provides some very useful and widely applicable scientific and technological outputs, it also creates something of a bias in the overall direction of biomass system development;[42] a bias towards high efficiency systems requiring high levels of technological and capital inputs and suitable for agricultural and climatic conditions very different from those that exist in the developing countries.

One of the important requirements, therefore, for maximising the potential for biomass energy in the developing world is the provision of an effective R and D effort which is directly aimed at local needs and conditions in these countries. This includes both basic research in a number of areas and adaptive research to modify existing concepts for practical application. No attempt is made here to define a list of research priorities. However, a summary is presented of some of the main areas where technical information is lacking and where well directed R and D could have an important impact. These include :

1. Identification and characterisation of the energy problems within developing countries and investigation into the long term environmental effects of present fuel collection practices.
2. Identification and breeding of plant species which are

adapted to the soil and climatic conditions of tropical countries, particularly in marginal areas and are suitable for energy purposes. This includes fast growing trees, starch and sugar crops, aquatic plants, oil and hydro-carbon plants and other species.

3. Investigation into optimum management techniques for energy crops in the context of likely limitations on the availability of fertilizer, water, land and capital and the requirement for maintaining food production.

4. Adaptation of conversion and end-use technologies to fit in with local energy needs and locally available skills, financing and resources.

5. Socio-economic research into patterns of energy use, the attitudes and needs of energy users and the problems of diffusion of biomass technologies.

It should be stressed that while the developed countries can clearly play an important role in assisting these R and D efforts the majority of this work can only be performed effectively within the developing countries. Indeed, this represents a very important component in the development of local planning, R and D, and implementation capabilities and can be combined effectively with personnel training programmes.

7.3 Opportunities for Collaboration

Practically every country today is actively involved in the formulation and development of energy policies and many are incorporating biomass options into these plans. There are also a large group of national and international agencies involved in energy matters. Often, however, there is a marked lack of communication and cooperation between these groups, which results in a significant duplication of effort and dilution of results. Even within countries and within organisations the lack of cooperation is often quite chronic.

In view of these problems, it is clear that there is considerable scope for improving collaboration between the groups involved in energy matters. This includes collaboration both within countries and organisations and between them and can take a variety of forms. Some of the major areas of potential collaboration include the following :

- Coordination of groups within developing countries involved in biomass energy development (universities, government ministries, philanthropic organisations, commercial enterprises, etc.).
- Communication between developing countries of technical data and information regarding implementation strategies and resulting problems and successes.
- Exchange of personnel between countries to improve training and dissemination of ideas.
- Coordination of the activities of international agencies involved in energy matters and better orchestration of their research, planning and assistance efforts.

One of the initial problems of coordinating energy activities, is the difficulty in finding out who is doing what at present, where they are doing it and how. In an attempt to assist in this matter, a tentative listing of organisations and institutions active in alternative energy work is presented in Appendix B.

7.4 Final Comments

The importance of traditional biomass fuels in the developing countries today is indisputable, as are the threats posed in many parts of the world by deforestation and inappropriate fuel collection practices. Without effective measures to curb deforestation and provide environmentally sound domestic energy alternatives the ecological and social consequences could be disastrous - as is already being experienced in a number of the most severely affected regions. The need to develop new and enlarged energy sources to reduce the burden of importing fossil fuels and permit future economic and social development, is also very clear.

The development of biomass energy has considerable potential in each of these important areas but as has been stressed throughout this report it is also subject to a variety of constraints and limitations, many of which are highly site-specific. In view of the wide range of factors which influence the suitability of biomass systems under local conditions, no general conclusions can be made at this stage regarding which biomass systems are most appropriate for the developing countries - or what their theoretical or practical impact might be. Countries must assess the potential for biomass systems on an individual basis and in the light of their own needs, their own development priorities and resource endowment, and in the context of their own socio-political and economic circumstances.

The notable success of a number of countries in implementing biomass schemes underlines what can be achieved. The problems encountered by other biomass projects also provide useful lessons as to what the likely barriers to implementation will be and an insight into how these might be avoided. For several well-tried biomass approaches - such as improved cooking stoves and charcoal kilns, biogas, community woodlots, and ethanol production from sugar cane - a substantial fund of practical experience has been built up on which further development work can be based (although there are certainly no prescriptions that can be written which will guarantee effective implementation of these schemes).

In addition to these well established biomass approaches, there are also a significant number of relatively new concepts which appear attractive both in technical and economic terms and are now being attempted in practice in several countries. These include large-scale tree plantations for electricity generation, the use of vegetable oils as liquid fuels, community biogas schemes and small-scale gasifiers for driving diesel engines. The results from these trials are eagerly awaited. Beyond that, the intense interest in biomass energy and the extensive research effort being carried out worldwide hold tremendous promise for the development of new and improved systems in the future.

Whatever the methods and technologies chosen, however, and regardless of theoretical advantages of particular systems, much can only be learnt by trial and error - and through first-hand experience in individual situations. Devising an appropriate policy towards biomass energy is undoubtedly a complex task - and providing for effective implementation of such a policy will be a challenge that will require imagination, innovation and a high level of commitment from the countries concerned - as well as full assistance from the developed nations. The era of cheap oil and abundant supplies of traditional fuels is over, however, for practically all developing countries. Coming to grips with the fundamental problems that energy supply poses and arriving at effective solutions is one of the most vital tasks in the overall development process. In this context, biomass energy has much to offer.

Acknowledgements

The authors are greatly indebted to Ainslee Rutledge for her skilled and diligent work in typing the manuscript and to Mr H.G. Barnard for his meticulous proof-reading and editing of the text. We would also like to acknowledge the helpful comments and suggestions on an earlier draft given by :

Mr Andrew Barnett : IDRC, Sussex University

Mr Jas Gill : Energy Research Group, The Open University

Dr Gerald Foley : IIED, London

Dr Leo Pyle : Imperial College, London.

Units and Conversion Factors

1 Gigajoule (GJ) = 10^9 joules

1GJ = 0.95 x 10^6 BTU

1GJ = 0.24 x 10^6 kilocalories

1GJ = 280 kWh

1 Tonne wood (air dry) = 15GJ

1 Tonne wood (oven dry) = 20GJ

1 Tonne agricultural residues = 13GJ

1 Tonne dung = 15GJ

1 Tonne charcoal = 30GJ

1 Tonne wood = 1.4m^3 wood

1m^3 wood = 10GJ

1 tonne oil = 44GJ

1 barrel oil = 6.3GJ

1 tonne coal = 28GJ

1 hectare = 2.47 acres

List of References

1. A. Agarwal (1979), 'Can biogas provide energy for India's rural poor?', Nature 281, 9.

2. Agency for International Development (1980), 'The Socio-Economic Context of Fuelwood Use in Small Rural Communities', AID Evaluation Special Study No.1

3. Alcohol production from biomass in the developing countries, (1980) World Bank, 1818 H St. N.W., Washington DC 20433, USA.

4. J.E.M. Arnold and J. Jongma (1978), Fuelwood and charcoal in developing countries, Unasylva 29, 2-9.

5. S. Aziz (1978), 'Rural Development : Learning from China', Macmillan, London.

6. G.W. Barnard and S. Ratasuk (1980), The potential for wood energy in Thailand, Thailand Inst. Sci. & Techn. Research, 196 Phahonyothin Road, Bankhen, Bangkok 9, 62pp.

7. A. Barnett, D.L. Pyle and S.K. Subramanian (1978), Biogas Technology in the Third World, IDRC, No.103e, Box 8500, Ottawa, Canada, 132pp.

8. M. Best (1979) The Scarcity of Domestic Energy : A Study in Three Villages, South Africa Labour and Development Research Unit, Working Paper No.27, Capetown.

9. L.P. Bharava and M.L.A. Sen (1970), Social Aspects of Forestry in the Arid Zone, Annals of Arid Zones, Vol.9, No.1, Arid Zone Research Association of India, Jodhpur.

10. Jan Bialy (1979), Measurement of the Energy Released in the Combustion of Fuels, Occasional papers on Appropriate Technology, ATO22, School of Engineering Science, University of Edinburgh, Scotland.

11. Bio-Energy Directory (1980), Bio-Energy Council, 1625 Eye St N.W., Washington DC, USA.

12. Bio Energy '80 Proceedings (1980), Bio-Energy Council, 1625 Eye St N.W., Washington DC, USA.

13. Biomass for Energy (1979), Proc. Conf. C-20, UK-ISES, 19 Albemarle St, London W1, UK, 99pp.

14. D. Bond (1980), State of the Art of Thermochemical Processes for Conversion of Biomass to Energy, Report prepared for United Nations technical panel on Biomass Energy, October 1980.

15. G.R. Breag and A.E. Chittenden (1979) Producer gas : its potential and application in developing countries, Tropical Products Institute G130, 56/62 Grays Inn Road, London, England.

16. J. Briscoe (1979), Energy Use and Social Structure in a Bangladesh Village, Population and Development Review 5, p.615.

17. N. Brown (1980), Renewable energy resources for the developing countries, Annual Review of Energy, 5, 389-413.

18. J. Burley (1978), Selection of Species for Fuelwood Plantations, 8th World Forestry Congress, Jakarta.

19. E. Celeski, J. Dunkerley and W. Ramsey (1979), Household energy and the poor in the third world, Research Paper R-15, Resources for the Future Inc., 1755 Mass Ave N.W., Washington DC.

20. T.I. Chiang et al (1976), Pyrolytic Conversion of Agricultural and Forestry Wastes in Ghana, Georgia Tech. and USAID Project A-1823, Report AID/ta-C-1290.

21. J.M. Cohen and N.T. Uphoff (1980), Participation's Place in Rural Development : Seeking Clarity through Specificity, World Development, Vol.8, 213.

22. Turi Hammer Digernes (1979), Fuelwood crisis causing unfortunate land use or the other way around, Norsk, Geogr. Tidsskr, Oslo, 33, 25.

23. J.S. Douglas and R.A. de J. Hart (1976), Forest Farming : Towards a solution to problems of World Hunger and Conservation, Robinson and Watkins Press, London.

24. G.S. Dutt (1978), Reducing Cooking Energy use in Rural India, Report PU/CES 74, Center for Environmental Studies, Princeton Univ., Princeton, November 1978.

25. D.E. Earl, (1975), Forest energy and economic development, Clarendon Press, Oxford, 128pp.

26. E. Eckholm, (1979), Planting for the future-forestry for human needs, Worldwatch Paper No.26, 1776 Massachusetts Ave N.W., Washington DC, USA, 64pp; also (1975) The other energy crisis : fuelwood, Worldwatch Paper No.1.

27. Energy for World Agriculture (1979), ed. B.A. Stout, FAO, via delle Terme di Caracalla, 00100, Rome, Italy.

28. Energy from Biological Processes, vols 1 and 2 (1980), Office of Technology Assessment, Washington DC 20510, USA.

29. Energy from Biomass (1980), Proc. Int. Conf., Brighton, England, P. Chartier et al (eds), Applied Science Publishers.

30. Energy in the developing countries (1980), World Bank, 1818 H St N.W., Washington DC, 20433, USA, 92pp.

31. Sabri Ergun (1980), Biomass Liquefaction Efforts in the United States, Lawrence Berkeley Laboratory Report, Berkeley California LBL-10456, February 1980.

32. G. Foley and A. van Buren (1980), Coal substitution and other approaches to easing the pressure on woodfueld resources - case studies in Senegal and Tanzania. International Institute for Environment and Development, London. Presented to 2nd Meeting of the technical panel on fuelwood and charcoal for the UN Conference on New and Renewable Sources of Energy.

33. D. French (1979), The economics of renewable energy systems for developing countries, US-AID, Washington DC 20523, USA, 67pp.

34. D. French and P. Larson, eds. (1980), Energy for Africa - selected readings, US-AID, Washington DC 20523, USA, 199pp.

35. R.C. Ghosh (1979), The socio-economic consequences and constraints to the use of land and forest for energy and organics in India, Proc. Int. Workshop on Biological and Sociological basis for a rational use of forest resources for energy and organics, Michigan State Univ., May 1979, S.G. Boyce (ed.), p.36.

36. D.O. Hall (1979), Solar energy use through biology - past, present and future, Solar Energy 22, 307-328; also (1979) Nature 278, 114-117.

37. D.O. Hall (1980), Regional development assisted through local renewable energy resources, Proc. Int. Conf. 'Natural Resources and Regional DEvelopment : the semi-arid regions case', Cocoyac, Mexico, Oct. 1980, Acad. Press, NY, E. Campos-Lopez, (ed) - in press.

38. D.O. Hall (1981), Solar energy through biology - fuel for the future, in J.T. Manassah and G.J. Briskey, eds., 'Advances in food producing systems for arid and semi-arid lands', Acad. Press, NY - in press.

39. D.O. Hall (1981), New Scientist 89, 1242, 524.

40. A.C. Harris (1978), Charcoal Production, Proc. 8th World Forestry Congress, Jakarta, Indonesia.

41. P. Harrison (1979), Inside the Third World, Penguin Books, England.

42. H.K. Hoffman (1980), Alternative Energy Technologies and Third World Rural Energy Needs : A case of Emerging Technological Dependency' Development and Change, July 1980.

43. Marilyn Hoskins (1980), Community participation in African Fuelwood Production, Transformation and Utilisation, in Energy for Africa - selected readings, edited by D. French and P. Larson, US-AID, Washington DC 20523, USA.

44. J.W. Howe and F.A. Gulick, eds (1980), Fuelwood and other renewable energies in Africa, Overseas Development Council, 1717 Massachusetts Ave N.W., Washington, DC 20636, USA, 98pp.

45. D.P. Hughart (1979), Prospects for traditional and non-conventional energy sources in developing countries, World Bank, Energy Department (WSP No.346), 1818 H St N.W., Washington DC 20433, 132pp.

46. R.J. de Lucia and M. Lesser (1980), Energy from organic residues in developing countries : present use, potentials and observations ob feasibility, Proc. Bio;Energy '80 (see below).

47. A. Makhijani and A. Poole (1975), Energy and agriculture in the Third World, Ballinger, Cambridge, USA, 168pp.

48. S.L. Meisel et al (1976), Chem. Tech. 6, 86.

49. J.C. Mello, C.W. Urban, Advantages of small and medium scale distilleries and their polyvalent character, Bio-Energy '80, p.403.

50. R.P. Morgan et al (1979), Appropriate technology for renewable resource utilisation, Chapter 4, Final report of AID Contract No. AID/DSAN/C/0061, Washington Univ., St Louis, Missouri, USA.

51. P.A. Moss (1981), Biomass in Developing Countries - Framework for Implementation, paper prepared for Technical Panel on Biomass Energy for UN Conference on New and Renewable Sources of Energy, University of London, King's College.

52. T.K. Moulik and Srivastava, U.K. (1975), Biogas plants at the village level : problems and prospects in Gujurat, Centre for Management in Agriculture Monograph 59, Indian Institute of Management, Ahmedabad.

53. P.K.R. Nair (1980), Agroforestry Species - A Crop sheets manual, ICRAF, P.O.Box 30677, Nairobi, Kenya.

54. National Academy of Sciences (1975), Leucaena. Promising Forage and Tree Crop for the Tropics, Washington DC.

55. National Academy of Sciences (1976), Making Aquatic weeds useful : some perspectives for Developing Countries, Washington DC.

56. K. Openshaw (1978), Woodfuel - a time for reassessment, Natural Resources Forum 3, 35-51.

57. P.F. Palmedo et al (1978), Energy needs, uses and resources in developing countries, BNL 50784, Brookhaven Nat'l Lab., Upton, NY 11973, USA, 143pp.

58. J.K. Parikh (1980), Energy systems and Development, Oxford University Press, Delhi, India, 152pp.

59. Pre-feasibility study of biogas technology application in rural areas of Thailand (1979), App. Sci. Research Corp. of Thailand, Submitted to IDRC, P.O.Box 8500, Ottawa, Canada, 127pp.

60. J.W. Purseglove (1968), Tropical Crops : Dicotyledons and Monocotyledons, Longmans, London.

61. D.L. Pyle (1977), Wood Pyrolysis - A working paper, Dept. Chemical Engineering, Imperial College, London.

62. D.L. Pyle (1979), Fuel alcohol production : some options and economics, and Biomass as a fuel source in the developing countries, ITDG Conference, Reading, England.

63. N.H. Ravindranath et al (1980), An Indian village agricultural ecosystem - case study of Ungra village, Proc. Conf. Energy from Biomass, Brighton, UK, November 1980, publ. by Applied Sci. Publ. London (in press).

64. A.K.N. Reddy and D.K. Subramanian (1979), The design of rural energy centres, Proc. Indian Acad. Sci., Vol.2, part 3, 395-416.

65. T.B. Reed (1980), The Combustion, Pyrolysis, Gasification and Liquefaction of Biomass, Proc. Conf., Energy from Biomass, Brighton, England.

66. R. Revelle (1976), Energy Use in Rural India, Science 192, 969.

67. R. Revelle (1979), Energy sources for rural development, Energy 4, 969-988; also (1980) Energy dilemmas in Asia, Science 209, 164-174.

68. R. Roy (1980, Family and Community Biogas Plants in Rural India and China, Alternative Technology Group, The Open University, England.

69. J.A. Semana (1979), Fuel Farming, paper presented at the Workshop on Energy from Biomass and Wastes, University of Peradeniya, Sri Lanka, 26-30 November 1979,

70. C.V. Seshadri et al (1978), Energy Plantations - a case study of the Coromandel Littoral, Shri A.M.M. Muragappa Chettiar Research Centre, Madras, India.

71. Dale V. Shaller (1979), Socio-cultural Assessment of the Lorena Stove and its diffusion in Highland Guatemala, Mimeo, VITA.

72. M. Slesser and C. Lewis (1979), Biological energy resources, Spon London, 192pp.

73. V. Smil and W.E. Knowland eds, (1980), Energy in the developing world : the real energy crisis, Oxford University Press, Oxford, 386pp.

74. John S. Spears (1978), The Changing Emphasis in World Bank Lending. A Summary of Recent Experiences and Problem Areas of Relevance to the Eighth World Forestry Contress, presented at the Eighth World Forestry Congress, Jakarta, Indonesia.

75. J.S. Spears (1980), Overcoming constraints to increased investment in forestry, Proc. 11th Commonwealth Forestry Conference, Trinidad, September 1980.

76. H.E.M. Stassen and T. Zijp (1980), The Gasification by Partial Combustion project in Tanzania, Progress Report July 1980, THT, P.O.Box 217, Enschede, The Netherlands.

77. A Survey of Biomass Gasification (1979), Solar Energy Research Institute, Golden, Colorado, 80401, USA.

78. S.K. Subramanian (1978), Biogas Systems in Asia : A survey, in Barnett et al (1978), op.cit.

79. R.P. Taylor (1980), The Peoples Republic of China : Experience and capabilities in renewable energy resource development, World Bank Science and Technology Report Series No.35, Washington DC, 20433.

80. C.E. Thibau (1978), Industrial Charcoal from Mixed Tropical Forests, Proc. 8th World Forestry Conference, Jakarta.

81. M.P. Todaro (1977), Economics for a developing world, Longmans, London, p.234.

82. S.C. Trindade (1980), Energy Crops - the case of Brazil, Proc. Int. Conf., Energy from Biomass, Brighton, England, November 1980. P. Chartier et al (ed) - in press.

83. Tropical Products Institute, The Production of Charcoal in a Portable Metal Kiln, G119, London.

84. E. Uhart (1976), The Wood Charcoal Industry in Africa, Memorandum for the fourth session, African Forestry Commission, FAO, AFC/76/10.

85. Ariane van Buren (1980), Biogas beyond China, First International Training Program for Developing Countries, Ambio 9, No.1, p.10.

86. W. Vergara and D. Pimental (1978), Fuels from biomass : comparative study of the potential in five countries, Adv, Energy Systems and Technology, Academic Press, NY, USA, vol.1, pp.125-173.

87. D.G. Wilson, Alternative Automobile Engines, Scientific American vol.239 , No.1.

88. World Energy Supplies 1970-1978 (1979), UN Department of International Economic and Social Affairs, New York.

89 B.R. Winterhalder, R. Larsen and R.B. Thomas, Dung as an Essential Resource in a Highland Peruvian Community, Human Ecology, 2.

General Texts Recommended for Further Reading

J. Dunkerley et al (1981), Energy Strategies for Developing Nations, Published for Resources for the Future by the Johns Hopkins University Press, Baltimore.

D.E. Earl (1975), Forest Energy and Economic Development, Clarendon Press, Oxford.

R.P. Moss and W.B. Morgan (1981), Fuelwood and Rural Energy, Production and Supply in the Humid Tropics, Natural Resources and the Environment Series, Vol.4, Published for the UN University by Tycooly International Publishing Ltd, 17 Gilford Road, Sandymount, Dublin 4, Ireland.

P.K.R. Nair (1980), ICRAF : Agroforestry Species : A Crop Sheets Manual, ICRAF, Box 30677, Nairobi, Kenya.

National Academy of Sciences, Washington, DC 20418:
Firewood Crops, Shrub and Tree Species for Energy Production.
Conversion of Tropical Moist Forests.

Palmedo, Nathans, Beardsworth and Hale (1978), Energy Needs, Uses and Resources in Developing Countries, National Center for Analysis of Energy Systems, Brookhaven National Laboratory, Associated Universities, Upton, NY 11973.

Smil and Knowland, Eds. (1980), Energy in the Developing World : the Real Energy Crisis, Oxford University Press, Oxford.

World Bank, 1818 H Street, NW Washington DC 20433 :
Energy in the Developing Countries.
Alcohol Production from Biomass in Developing Countries.
Energy Options and Policy Issues in Developing Countries.

Worldwatch Institute, 1776 Massachusetts Ave N.W., Washington D.C., USA:
Worldwatch Paper 35 : Food or Fuel : New Competition for the World's Cropland.
Worldwatch Paper 42 : Wood : An Ancient Fuel with a New Future.

Appendix A

Data on Biomass Energy Use

PART 1

VILLAGE ENERGY STUDIES*

Bangladesh Dhanishwar Village (Bangladesh Energy Study,1978, and Hughart,1979)

422 people (77 families)
Total Village demand for fuel for cooking = 2391 GJ/year
= 5.7 GJ/person/year
= .4t/person/year
Rice straw provides 75% cooking fuel
Jute sticks provide 15% cooking fuel
Fuelwood provides 7% cooking fuel

(Briscoe, 1979) Ulipur Crop residues = 54% total energy use
Dung = 2.7% total energy use
Firewood = 20% total energy use
Other vegetable matter = 17.7% total energy use

Total energy use = 1,615 10^3Kcal/person/year
= .5t/person/year
= 7.5 GJ/person/year

Bolivia (Makijani and Poole, 1975) Quebrada Household
Fuelwood use = 2t/person/year = 30 GJ/person/year
= 100% of energy use

Botswana (White, 1979) Matsheng Villages
Firewood use = 1.1t/person/year
= 16.5 GJ/person/year

Cameroon (Vennetier, 1979) Ngaoundere

Wood fuel - average use is 1.6-1.7 m^3/person/year
= 1.2-1.25t/person/year
= 18 GJ/person/year

Chad (Bertrand, 1977) N'Djamena

Woodfuel use, urban area = 4 m^3/person/year
= 2.9t/person/year
= 44 GJ/person/year

China (Makijani and Poole, 1975) Peipan Village

Biomass energy use = 21 GJ/person/year
= 1.5t/person/year
Biomass energy = 87% of energy use

Congo, People's Republic (Gilbert, 1978) Brazzaville

Woodfuel use = 1.22 m^3/person/year
= .9t/person/year
= 13.5 GJ/person/year

India (Makijani and Poole, 1975) Mangoan Village
Biomass energy use = 4.1 GJ/person/year
= .3t wood/person/year
This was 99% of energy use

* Detailed references are given on pp.201-205.

<u>Togo</u> (Grut, 1971) <u>Lome</u>

175 kg firewood plus 82 kg charcoal/person/year
(.18t or 2.7 GJ) (.08t or 2.5 GJ)
Total consumption = 5.2 GJ/person/year

<u>Upper Volta</u> (Ernst, 1978) <u>Village of Ranga</u>
234 people
Wood consumption = 0.84 kg/person/year
= 0.3t/person/year
= 4.5 GJ/person/year
Millet stalks (six months use per year) = 4.5 GJ/person/year
= 0.3t/person/year
Combined use = 0.6t wood/person/year

<u>Village of Boulenga</u>

Wood consumption = 1.49 kg/person/day
= 0.5t/person/year
= 8.2 GJ/person/year
Millet stalk consumption = 0.8 kg/person/day
= 0.3t/person/year
= 4 GJ/person/year
Combined use = 0.8t/person/year

(Bertrand, 1977) <u>Ouagadougou</u>
Wood use = 2.4 m^3/person/year (including building materials)
= 1.8t/person/year
= 27 GJ/person/year

Rwanda (Gatera, 1978) Nyarugenge, suburb of Kigali

81% of population used only charcoal as fuel
Charcoal use: 377 kg/person/year
If 12t wood - produce 1t charcoal then .38t charcoal = 4.5t wood/person/yr

Senegal (Tall, 1974) Dakar

Charcoal use = 117 kg/person/year = .12t/person/year = 3.5 GJ/person/yr
Charcoal used exclusively, due to distance of source- transport a problem

(Bertrand, 1977) Dakar

Wood use = 2 m^3/person/year, including building materials
= 1.5t/person/year
= 22 GJ/person/year

Sierra Leone (Cline-Cole, 1979) Waterloo, small village

Wood use = 1160 kg/person/year
= 1.2t/person/year
= 18 GJ/person/year

South Africa (Best, 1979)

Jozanna's Nek, Transkei Wood and dung use = .35t/person/year
= 5.3 GJ/person/year
This was 95% of total energy use
Mashunka, Kwazulu Wood use = 1.1t/person/year
= 17GJ/person/year; this was 99% of total energy

Sri Lanka (Bialy, 1979) Village near Anuradhapura

203 people in 34 households
Average firewood consumption for family of six = 50 kg/week
Therefore, in village(poor), firewood use = .35t/person/year
Compares this with urban consumption of .35t/person/year
and consumption in most villages of .52t/person/year

Sudan (Digernes, 1979) Bara Village

About 10,000 persons
Since mid-1978 people have abandoned kerosene stoves and refrigerators
Since early 1977 oil for households only available on black market
Roundwood consumption = 4.9 m^3/person/year as charcaol
and + 0.8 m^3/person/year as fuelwood
Total consumption 5.7 m^3/person/year
5.7 m^3 roundwood = 4.1t wood/person/year
Wood energy > 90% of total energy

(FRIDA, 1980) Khartoum

Firewood comsumption = 1.29 m^3/person/year = .9t/person/year = 13.5GJ
Charcoal consumption = 95 kg/person/year = .95t/person/year = 28.5 GJ

(Mukhtar, 1978) Khartoum
Charcoal consumption = 89.9kg/person/year = .9 t/person/year = 27 GJ
Many rely only on charcoal

Tanzania (Makijani and Poole, 1975) Kilombero Village in North

Biomass energy = 1.5t wood/person/year = 22.5 GJ/person/year

Mauritania (FRIDA,1980) For very desertic region near Nouakchott

Total firewood consumption = 2.13 m^3/person/year = 1.5t/person/year
= 22.5 GJ/person/year

Total charcoal consumption = 49 kg/person/year = .05t/person/year
= 1.5 GJ/person/year

Mexico (Makijani and Poole, 1975) Arango Village
420 people in 80 households
Biomass (residue) use = 13 GJ/person/year
= 0.9t wood/person/year
= 33% of energy use

Nepal (Hughart, 1979) Typical Hill Village

250 people in 48 households
Fuelwood use = 2 kg/person/day
= 0.73t/person/year

(Earl, 1975) Bardia District

Fuelwood use = 1 m^3/person/year
= 0.73t/person/year

Niger (Pare, 1979) Niamy

Wood use = 1 m^3/person/year
= 0.73t/person/year
= 11 GJ/person/year

(Delwaulle and Roederer, 1973) Niamy

Average wood use = 1.15 m^3/person/year
= 0.84t/person/year
= 12.6 GJ/person/year

Nigeria (Afolabi Ojo, 1979) Osu Village

90% of people used fuelwood even though the preference was for kerosene and electricity

(Makijani and Poole, 1975) Batagawara Village in N. Nigeria
1,400 people in 170 households
Biomass energy = 16 GJ = 1t/person/year
Biomass = 99% of total energy

(Grut, 1973) Firewood use in eight Nigerian cities

Ilorin Firewood use = 2.1m^3/person/year = 1.53t/person/year
Jos Firewood use = 1.6 m^3/person/year = 1.17t/person/year
Kaduna Firewood use = 2.3 m^3/person/year = 1.68t /person/year
Kano Firewood use = 1.7 m^3/person/year = 1.24t/person/year
Katsina Firewood use = 2.5 m^3/person/year = 1.83t/person/year
Maidugari Firewood use = 2.2 m^3/person/year = 1.6t/person/year
Sokotu Firewood use = 1.6 m^3/person/year = 1.2t /person/year
Zaria Firewood use = 1.9 m^3/person/year = 1.4t/person/year

(Ay, 1978) Ibadan Area

Fuelwood use: Village = 1.1m^3/person/year = .8t/person/year
Fuelwood use: City = .9 m^3/person/year = .66t/person/year

Peru (Winterhalder, Larsen, Thomas, 1974) Nunoa, Peruvian highlands

87% of families use dung as fuel
Dung use = 11t/family/year

<u>India</u> (cont.)

(Murugappa Chettiar Research Centre, 1981) <u>Injambakkam Village</u>

944 people in 211 households
Firewood use = 27.5 kg/person/month
= .33t/person/year
95% use firewood, 3% use only kerosene, and 1% use only dung

(Ravindranath, 1980) <u>Ungra Village</u>

932 people in 149 households
95% of total energy supplied by firewood and agro-waste
Firewood and agro-waste use = 0.6t/person/year = 9 GJ/person/year

(Reddy, 1979) <u>Pura Village</u>

357 people in 56 households (similar in size to 60% of India's villages)
89% of energy from firewood
2% of energy from kerosene
1% of energy from electricity
1% of energy from bullocks
7% of energy from humans

Energy use-91% domestic
-4% industry
-3% agriculture (77% from humans and bullocks)
-2% lighting
Human energy---- 2.3 hours/day collecting 10 kg wood per household
Firewood use = 0.6 t/person/year = 9 GJ/person/year

<u>Iran</u> (Vojdani, 1978) <u>Village in Semnan, Kavir</u>

150 people
Firewood use = 180t/year
= 1.2t/person/year
= 18 GJ/person/year

<u>Kenya</u> (Mutula Mun'gala, 1979) <u>Machakos District</u>

Average Wood and charcoal use = 1.5 m^3/person/year = 1.1t/person/year
Average rural household = 1.29 m^3/person/year = .9t/person/year
Average urban household = 1.79 m^3/person/year = 1.3t/person/year

<u>Lesotho</u> (Best, 1979) <u>Malefiloane</u>

Wood and dung use = .55t/person/year = 8.2 GJ/person/year
This was 98% of total energy use

<u>Mali</u> (Bertram, 1977) <u>Bamako</u>

Wood use = 1.5t/person/year = 23 GJ/person/year

(Sankare, 1979) <u>Bamako</u>
Average total consumption of 370 kg firewood = .37t/person/year
= 5.5 GJ/person/year
<u>plus</u> 17kg charcoal/person/year = .017t/person/year
= .5 GJ/person/year

(Caude, 1977) Wood Consumption in two small towns:
<u>Deguela</u> : 485-525 kg/person/year = .49t-.53t/person/year
= 7.3-7.9 GJ/person/year
<u>Sanzana</u> : 800-927 kg/person/year = .8-.9t/person/year
= 12-13.9 GJ/person/year

TRADITIONAL FUEL CONSUMPTION IN THE VILLAGE SECTOR - SUMMARY

Country	Village	Number of persons, families or households	Energy Source	% of total from biomass	Fuel use (per person per year) GJ	Tons equivalent wood
Bangladesh	Dhanishwar	77 families, 422 people	Rice straw, fuelwood	100	6 GJ	.4t
Bangladesh	Ulipur	48 families, 330 people	Crop residues plus firewood	100	7.5 GJ	.5t
Bolivia	Quebrada	6 people	Fuelwood	100	30 GJ	2t
Botswana	Matsheng		Firewood		16.5 GJ	1.1t
Cameroon	Ngaoundere		Firewood		18 GJ	1.2-1.25t
Chad	N'Djamena		Firewood		44 GJ	2.9t
China	Peipan	1000 people, 200 households	Mostly firewood	87	21 GJ	1.5t
Congo, People's Republic	Brazzaville		Firewood		13.5 GJ	.9t
Ethiopia	Addis Ababa		Fuelwood		10.5 GJ	.7t
India	Pura	357 people, 56 households	Mostly wood	96	9GJ	.6t
India	Injambakkam	944 people, 211 households	Firewood	95	5 GJ	.33t
India	Mangoan	1000 people	Dung, residues	99	4.1 GJ	.3t
India	Ungra	932 people, 149 households	Fuelwood, residues	95	9 GJ	.6t
Iran	Semnan, Kavir	150 people	Firewood		18 GJ	1.2t
Kenya	Machakos District		Wood and charcoal		15 GJ	1t
Lesotho	Malefiloane		Wood, dung	98	8.2 GJ	.55t
Mali	Deguela		Wood		7.3-7.9 GJ	.49-.53t
Mali	Sanzana		Wood		12 - 14 GJ	.8-.9t
Mali	Bamako		Wood		22.5 GJ	1.5t
Mauritania	Nouakchott		Wood		22.5 GJ	1.5t
Mexico	Arango	420 people, 80 households	Residues	33	13 GJ	0.9t
Nepal	Hill	250 people 48 households	Fuelwood	97	11 GJ	.73t
Niger	Niamy		Fuelwood		12.6 GJ	.84t
Nigeria	Batagawara	14,000 people, 170 households	Biomass	99	15 GJ	1t

SUMMARY (continued)

Country	Village	Number of persons, families or households	Energy Source	% of total from biomass	Fuel use (per person per year) GJ	Tons equivalent wood
Nigeria	Kano	348,000	Firewood	*	18 GJ	1.2t
Nigeria	Ilorin	246,000	Firewood	*	22.5 GJ	1.5t
Nigeria	Zaria	196,000	Firewood	*	21 GJ	1.4t
Nigeria	Kaduna	177,000	Firewood	*	25.5 GJ	1.7t
Nigeria	Maidugari	165,000	Firewood	*	24 GJ	1.6t
Nigeria	Katsina	107,000	Firewood	*	27 GJ	1.8t
Nigeria	Jos	107,000	Firewood	*	18 GJ	1.2t
Nigeria	Sokoto	106,000	Firewood	*	24 GJ	1.2t
Nigeria	Ibadan				12 GJ	.8t
Peru	Nunoa	15 families studied	Dung	87	165GJ/fam./yr.	11t/fam./yr.
Rwanda	Nyarugenge		Charcoal	81	51 GJ	3.4GJ
Senegal	Dakar		Charcoal		22 GJ	1.5t
Sierra Leone	Waterloo	"Small Village"	Wood		18 GJ	1.2t
South Africa	Jozanna's Nek		Wood and dung	95	5.3 GJ	.35t
South Africa	Mashunka		Wood	99	17 GJ	1.1t
Sri Lanka	village nr. Anuradhapura	203 people in 34 households	Firewood		5.3 GJ	.35t
Sudan	Bara	10,000 people 1600 households	Mostly charcoal	90+	62 GJ	4.1t
Sudan	Khartoum		Charcoal	Many 100%	27 GJ	1.8t
Tanzania	Kilombero	100 people	Biomass	100	23 GJ	1.5t
Togo	Lome		Wood and charcoal		5.5 GJ	.4t
Upper Volta	Ranga	234 people	Wood, millet stalks		9 GJ	.6t
Upper Volta	Boulenga		Wood, millet stalks		12 GJ	.8t
Upper Volta	Ouagadougou		Wood		27 GJ	1.8t

*(Charcoal and cornstalks also used but use not measured in this study)

PART 2

NATIONAL ENERGY STUDIES*

Algeria (Pyle, 1980) Biomass use is .05 times commercial energy use = 1.1 GJ/person/year = 0.07t/person/year

(Frisch, 1980) 1975: .42 million tons wood used.
This was 2.3% of the total energy used.

(Howe and Gulick, 1980)

1976: Wood fuel use = .066t/person/year = 1 GJ/person/year
Wood fuel = 4% of total energy consumption

Angola (Howe and Gulick, 1980)

1976: Wood fuel consumption = 1t/person/year = 15 GJ/person/year
Wood fuel = 74% of total energy use

Bangladesh (FAO, 1976)

Fuelwood and charcoal use = 1.9 GJ/person/year = 0.1t/person/year
70% of national energy use in form of traditional fuels

(Commonwealth Science Council, 1978)

75% of energy use derived from biomass = 3 GJ/person/year
Biomass used - 25% dung
- 42% rice residues
- 6% jute sticks
- 4% firewood
- 6% bagasse

Barbados (Cox, 1978)

Bagasse use = 12 GJ/person/year
= 0.8t wood/person/year
Bagasse supplies 29% of primary energy supply

Benin (Digernes, 1979)

Fuelwood use = $9.11 \times 10^6 m^3$/year
= 2.9 - 3.5 m^3/person/year
= 2.1 - 2.6t/person/year
= 28 - 34 GJ/person/year

(Howe and Gulick, 1980)

1976: Woodfuel use = .66t/person/year = 10 GJ/person/year
Woodfuel = 86% of total energy use

(FAO, 1976)

Fuelwood and charcoal use = 8.2 GJ/person/year
= 0.6t/person/year

Bolivia (FAO, 1976)

Fuelwood and charcoal use = 8.3 GJ/person/year = 0.6t/person/year

Botswana (Thipe and Mokobi, 1979)

Firewood use = 0.75 - 1t/person/year
= 11 - 15 GJ/person/year
70% of firewood used for heating water

* Detailed references are given on pp.201-205.

Botswana(cont.)

(FAO, 1976)

Fuelwood and charcoal use = 10.6 GJ/person/year = 0.7t/person/year

Brazil (Frisch, 1980)

1975: Wood use = 54 million tonnes
Wood and agricultural waste use = 29% of total energy use

(Muthoo, 1978) 1976 data

Fuelwood 2.13×10^7 tce ($118 \times 10^6 m^3$) = 21.5% of energy use
Charcoal 3.15×10^6tce ($24 \times 10^6 m^3$) = 3.2% of energy use
Bagasse 4.17×10^6 tce = 4.2% of energy use
Total biomass = 2.87×10^7tce = 1.28×10^{18}J = 11.7 GJ/person/year
= 0.8t wood/person/year

This was 29% of total energy

(Vergara and Pimentel, 1978)

Biomass provides 31% of total energy use
Biomass provides 53% of national energy production
Biomass energy produces = 207×10^{12}Kcal = 8.7×10^{17}J
= 7.9 GJ/person/year

(Suarez, 1979)
Rural areas - 92% of domestic energy requirements come from biomass

(Gochnarg, 1979)
In 1977 oil imports provided 41% of total energy use and cost
\$4.5 billion (⅓ total value of imports). In 1979 oil imports
will cost \$6 billion = ½ total value of exports.

(FAO,1976)
Fuelwood and charcoal use = 13.8 GJ/person/year
= 0.9t/person/year

Burundi (Howe and Gulick, 1980)
1976: Woodfuel use: .2t/person/year = 3.2 GJ/person/year
Woodfuel = 89% total energy use

Cameroon (Howe and Gulick, 1980)
1976: Woodfuel use: .96t/person/year = 14.5 GJ/person/year
Woodfuel = 82% total energy use

Central African Empire (Howe and Gulick, 1980)
1976: Woodfuel comsumption = 1t/person/year = 15 GJ/person/year
Woodfuel = 91% of total energy use

Chad (Howe and Gulick,1980)
1976: Woodfuel use = .77t /person/year = 11.5 GJ/person/year
Woodfuel = 94% of total energy use

Chile (Frisch, 1980)
1975: Woodfuel use = 3.3 million t wood
This was 13.3% total energy use

China (Chen, 1980)
Total biomass use = 8.35×10^{18}J/year
Population = 1×10^9 people
But rural population = 0.8×10^9 people
Each person = approx. 10 GJ/year = ⅔t/person/year
Potential biomass = 11.45×10^{18}J/year

China (Cont'd)

(Smil, 1978)

Forest fuel use = 1.4 x $10^8 m^3$ = 1.4 x 10^{18} J/year = 1.6 GJ/person/year
(15% of forest increment harvested annually = 1.596 x 10^{18} J)
Crop residues used = 1.75 x 10^8 t/year (conservative estimate)
= 7 x 10^{14} kcal = 2.9 x 10^{18} J/year
= 3.2 GJ/person/year

(FAO,1976)

Fuelwood use = 1.7 GJ/person/year
Wood and animal and vegetable wastes = 12GJ/person/year
equivalent to 0.9t wood/person/year

Congo (Howe and Gulick, 1980)

1976: Woodfuel use = 1.3t/person/year = 18.7 GJ/person/year
Woodfuel use = 80% total energy use

Costa Rica (Frisch, 1980)

1975: Wood use = 1.4 million t wood, or 34% total energy use

(SIECA & CEPAL, 1978)

28% of total energy consumption was fuelwood and charcoal
14% of total energy consumption was bagasse
66% of households used fuelwood for cooking

Egypt (Pyle, 1980)

Biomass use is .35 times commercial energy use
= 4.9 GJ/person/year = 0.33t/person/year

(Frisch, 1980)

1975: Wood and agricultural waste accounted for 6.6% of total energy use

(Ibrahim, 1979)

Crop residues = 12% total energy use
Dung = 2% total energy use
Wood = 0.48% total energy use
Total non-commercial energy = 14.5% of total energy use

El Salvador (SIECA & CEPAL, 1978)

1975: 37% of total energy consumption was fuelwood and charcoal
1975: 9% of total energy consumption was bagasse
1978:77% of households used fuelwood for cooking

Ethiopia (FRIDA, 1980)

Fuelwood consumption = 0.94 m^3/person/year in vicinity of Addis Ababa
= .7t/person/year = 10 GJ/person/year

(Uhart, 1975)

Charcoal consumption = 90 kg/person/year = .09t/person/year

(Howe and Gulick, 1980)

1976: Woodfuel use = .78t/person/year = 11.7 GJ/person/year
Woodfuel = 93% of total energy

Gabon (Howe and Gulick, 1980)

1976: Woodfuel use = 2.2t/person/year = 34 GJ/person/year
Woodfuel use = 44% of total energy use

Gambia (Openshaw, 1976)

Fuelwood = 1.2t/person/year
Fuelwood energy = 62% of total energy

Gambia (Cont'd)

(Howe and Gulick, 1980)
Woodfuel use = .46t/person/year = 7 GJ/person/year
Woodfuel = 73% of total energy use

(Floor, 1977)
Woodfuel is 62% of total energy consumption.
Per capita wood consumption = 1.2 m^3/year = 0.9t/person/year
= 13.5 GJ/person/year
Estimated fuelwood consumption is 2.3 x $10^5 m^3$ for rural areas (85% of population) and 4 x $10^4 m^3$ for urban areas

(Arnold and Jongma, 1978)
Average woodfuel use = 1.6 m^3/person/year = 1.1t/person/year
= 16.5 GJ/person/year

(FAO, 1976)
Fuelwood and charcoal use = 4.9 GJ/person/year
= 0.3t/person/year

(Gambia Land Resources Project, 1973)
1972: Total wood and charcoal consumption = 1.2t/person/year countrywide
Countrywide, 99% used wood and charcoal
Rural population - woodfuel consumption was 1.1t/person/year
100% used woodfuel
Urban population - woodfuel consumption was 1.25t/person/year
96% used woodfuel

Ghana (Howe and Gulick, 1980)
1976: Woodfuel use = 1t/person/year = 15 GJ/person/year
Woodfuel use = 74% of total energy use

Guatemala (SIECA & CEPAL, 1978)
1975: 41% of total energy consumption was fuelwood and charcoal
1975: 7% of total energy consumption was bagasse
1978: 88% of households used fuelwood for cooking

Guinea (Howe and Gulick, 1980)
1976: Woodfuel use = .58t/person/year = 8.7 GJ/person/year
Woodfuel use = 74% of total energy use

Guinea Buisseau (Howe and Gulick, 1980)
1976: Woodfuel use = .82t/person/year = 12 GJ/person/year
Woodfuel use = 87% of total energy use

Honduras (SIECA & CEPAL, 1978)
1975: 45% of total energy consumption was fuelwood and charcoal
1975: 6% of total energy consumption was bagasse
1978: 81% of households used firewood for cooking

India (Frisch, 1980)
Wood and agricultural waste provided 57% of total energy use

(Vergara and Pimentel, 1978)
Biomass use = 0.65t/person/year = 10 GJ/person/year
Biomass = 40% of total energy use and 68% of national energy production

(Ramachandran and Gururaja, 1978)
Biomass provides 52% of total energy use
(Wood: 27% Dung: 15% Wastes: 10%)

India (Cont'd)

(Ghosh, 1979)

Wood is the principal source of energy in India - rural people (80%) and urban poor

Annual domestic energy consumption = 0.35 tce/person/year
= $1.05m^3$/person/year
= .8t/person/year
= 11.5 GJ/person/year

(60% from wood and 31% from residues and dung)

203 m^3 (82 m tce or 1.69×10^{18} J) fuelwood consumed each year (is 90% of wood used each year) - deficit of 100 m^3 in 1990

(Argal, 1978)

Of total energy used (235m tce) yearly, 52% or 121m tce is from biomass

Per capita wood use = 0.4t/person/year = 6 GJ/person/year

Fuelwood = 74m tce(32%); Dung = 30m tce; Wastes = 17m tce

Domestic energy use - half of energy used in country
- 92% from non-commercial energy
- 70% from fuelwood

(Arnold and Jongma, 1978)

Woodfuel = 34% of total energy use

Biomass = 56% of total energy use

Biomass = 93% of rural domestic energy use

(Revelle, 1979)

Rural India - annual energy use = 0.35t coal/person (89% is non-commercial)

All India - annual energy use = 0.49t coal/person (0.23t or 47% is from biomass)

Iran (FAO,1976)

Fuelwood, residues and dung = 14.9 GJ/person/year = 1t/person/year

Fuelwood and charcoal use = 2.1 GJ/person/year = 0.1t/person/year

Iraq (Pyle, 1980)

Biomass energy use = 0.2 GJ/person/year = 0.01t/person/year

Ivory Coast (Howe and Gulick, 1980)

1976: Woodfuel use = .7t/person/year = 11 GJ/person/year
Wood use = 46% of total energy use

Jamaica (1977 study)

Bagasse equivalent to 1.9×10^6 barrels of oil/year
= 5.4 GJ/person/year
= 0.4t wood/person/year

Bagasse supplies 29% of local energy use (excluding bauxite industry)

Jordan (Pyle, 1980)

Biomass energy use = .1 GJ/person/year = less than 0.01t/person/year

Kenya (Nyoike, 1979)

Fuelwood provides cooking and heating energy requirements for 90% of population

Consumption is greater than 10×10^6t/year
= 0.7t/person/year
= 10.5 GJ/person/year

(Openshaw, 1979)

Average fuelwood use = 1.5 m^3/person/year = 1.1t/person/year = 16.5 GJ

Household of six would have to pay $120/year if bought = slightly less than paraffin and ⅔ cost of gas

Kenya (Cont'd)

(Kabagambe, 1976)
Urban charcoal use = 0.1 - 0.17t/person/year = 3 - 5 GJ/person/year
80% of urban people rely predominantly on charcoal

(Arnold and Jongma, 1978)
Average woodfuel use (1960) = 1 m^3/person/year
= 10 GJ/person/year

(Digernes, 1979)
1.2 to 1.5 m^3/person is the annual fuelwood use
.9 - 1.1t/person/year = 13.5 - 16.5 GJ /person/year

(Spurgeon, 1979)
Annual fuelwood consumption = 20 x 10^6t = 1.5t/person/year = 22.5 GJ
Rural economy - nearly 100% wood and charcoal

(Marguand and Githinji, 1979)
Fuelwood and charcoal = 70 - 80% of total energy use
Fuelwood use = 1.5 x $10^6 m^3$ = 1.1 m^3/person/year = .8t/person/year
= 12 GJ/person/year

(Muchiri, 1978)
Firewood use = 4.3 x 10^6tce = 10 GJ/person/year = 1 m^3/person/year

(Kokwaro, 1979)
Rural family (say seven) uses 1.5 - 2.3 m^3/month of firewood
= 2.5 - 3.9 m^3/person/year = 25 - 39 GJ/person/year
= 1.8 - 2.8t/person/year

(Western and Ssemakula, 1979)
Wood use - between 1 (low) and 2 (high) m^3/person/year
= 4.5 - 8.5 x 10^6tce = 73 - 84% of total energy use

(FAO, 1976)
Fuelwood and charcoal use = 9.1 GJ/person/year = 0.6t/person/year

(Howe and Gulick, 1980)
1976: Woodfuel use = .95t/person/year = 14 GJ/person/year
Woodfuel = 74% of total energy use

Lebanon (Pyle, 1980)
Biomass energy use = 0.2 GJ/person/year = 0.01t/person/year

Liberia (Howe and Gulick, 1980)
1976: Woodfuel use = 1.1t/person/year = 16 GJ/person/year
Wood use = 53% of total energy use

Libya (Howe and Gulick, 1980)
1976: Woodfuel use = .2t/person/year = 3 GJ/person/year
Woodfuel use = 5% of total energy use

(Pyle, 1980)
Biomass use = 1.9 GJ/person/year = 0.13t/person/year

Madagascar (Howe and Gulick, 1980)
1976: Woodfuel use = .64t/person/year = 9.6 GJ/person/year
Woodfuel use = 80% of total energy use

Malaysia (Frisch, 1980)
1975: Wood use = 1.12 million t wood. This was 8% of total energy use

Malawi (Howe and Gulick, 1980)
1976: Woodfuel use = .6t/person/year = 8.7 GJ/person/year
Woodfuel use = 82% total energy use

(FAO,World Bank, 1978) Malawi Wood Energy Project
Fuelwood consumption in 1977 = 8.85 million m^3, plus building poles 1.5 million m^3; 51% fuelwood use was in rural households
44% of fuelwood consumption was used in tobacco curing

(Jackson, 1980)
92% of total population lives in rural areas and is dependent on woodfuel for cooking and heating; domestic use = 0.6t/person/year or 49% total
Tobacco curing in 1980 used 4,667,100 m^3 fuelwood or 46.6% total demand
If include industrial use of firewood, then per caput use = 1.8 m^3 = 1.3 t = 19.5 GJ

Mali (Howe and Gulick, 1980)
1976: Woodfuel use = 2.1t/person/year = 32 GJ/person/year
Woodfuel use = 97% of total energy use

(Service for Water and Forestry, 1980)
Total average consumption was 657 kg firewood plus 182 kg charcoal
.66t wood = 9.9 GJ .18t charcoal = 5.5 GJ
Consumption ranged from 584 kg to 912 kg (firewood only)

(FRIDA, 1980)
Firewood consumption = 1.17 m^3/person/year
= 819 kg/person/year = 12.3 GJ/person/year
Charcoal consumption = 150 kg/person/year = .15t/person/year
= 4.5 GJ/person/year

(Bertrand, 1979)
Charcoal consumption was 495 g/person/day = 181 kg/person/year for families that depend exclusively on charcoal
.18t/person/year = 5.4 GJ/person/year

Mauritania (Howe and Gulick, 1980)
1976: Woodfuel use = .4t/person/year = 6 GJ/person/year
Woodfuel use = 63% of total energy use

(Pyle, 1980)
Biomass use is 1.5 times commercial energy use = 4.5 GJ/person/year
= 0.3t/person/year

Mauritius (Howe and Gulick, 1980)
1976: Woodfuel use = .02t/person/year = .3 GJ/person/year
Woodfuel use = 2% total energy use

Mexico (Frisch, 1980)
1975: Wood and agro-wastes accounted for 9% of total energy use

(FAO, 1976)
Fuelwood and charcoal use: 1.5 GJ/person/year = 0.1t/person/year

Morocco (Howe and Gulick,1980)
1976: Woodfuel use = .2t/person/year = 2.3 GJ/person/year
Woodfuel use = 19% of total energy use

(Pyle, 1980)
Biomass use is .21 times commercial energy use = 1.7 GJ/person/year
= 0.11t/person/year

Mozambique (Howe and Gulick, 1980)
1976: Woodfuel use = .85t/person/year = 12.8 GJ/person/year
Woodfuel use = 74% of total energy use

Nepal (Joshy, 1979)
Domestic sector consumes 93% of energy which is 99% non-commercial (<1% from kerosene and electricity)
Non-commercial energy - 97% fuelwood, 1% husks, 1% vegetable wastes, and .5% dung.
6.2×10^6 tce wood/year = 12.4×10^6t wood/year
= 0.9t wood/person/year = 13.5 GJ/person/year

(FAO , 1976)
Firewood and charcoal use = 7.4 GJ/person/year = 0.5t/person/year

(Earl, 1975)
Average fuelwood use = 0.25tce/person/year = 0.43t wood/person/year
92% of energy use is non-commercial

(Donovan, 1981)
Average fuelwood use = 1.4 m^3/person/year
Assumes the weight of one cubic meter of firewood to be 600 kilograms, thus, 1.4 m^3 = 0.85t

Nicaragua (SIECA & CEPAL, 1978)
1975: 25% of total energy consumption was fuelwood and charcoal
1975: 12% of total energy consumption was bagasse
1978: 75% of households used fuelwood for cooking

Niger (Howe and Gulick, 1980)
1976: Woodfuel use = .5t/person/year = 8GJ/person/year
Wood use is 87% of total energy use

(Raynaut, 1980) Maradi Province
Wood and grain stems use = .2t/person/year = 3GJ/person/year
½ wood and twigs and ½ stems of milo and sorghum

(FRIDA, 1980)
Fuelwood consumption = 0.47 m^3 = 329 kg/person/year = .33t/person/year
Charcoal consumption = 64 kg/person/year = .64t/person/year

(Castan, 1966)
Fuelwood consumption = 523 kg/person/year in rural areas

Nigeria (Afolabi Ojo, 1979)
Wood energy = 91% of total energy use

(Howe and Gulick, 1980)
1976: Woodfuel use = 1t/person/year = 15 GJ/person/year
Woodfuel use = 82% of total energy use

(McComb and Jackson, 1970)
Fuelwood consumption = 1.36 m^3/person/year = 1t/person/year

Panama Republic (SIECA & CEPAL, 1978)
1975: 13% of total energy consumption was fuelwood and charcoal
1975: 6% of total energy consumption was bagasse
1970: Overall, 40% of households used firewood for cooking (75% in rural areas, 5% in urban areas)

Papua New Guinea (Ministry of Energy, 1978)
Sinasina region - firewood use = 0.4t/person/year = 6 GJ/person/year
National average = 0.25t/person/year = 3.8 GJ/person/year

(Cont'd)

Papua New Guinea (Ministry of Energy, 1978) (Cont'd)
Conservative estimate - firewood = 39% of total energy demand
90% of firewood use is in village communities which contain 80% of the population (rural electrification is non-existent)
Firewood provides 5% of energy use in industry
Coffee husks provide 0.6% of energy use in industry
Charcoal production = 30t per month, all used in tea industry
Urban household consumption is 48% firewood

Peru (Hughart, 1979)
Wood use = 1.2t wood/person/year = 18 GJ/person/year in rural areas which contain 50 to 60% of the population
National average of firewood and charcoal use = 9.2 GJ/person/year = 0.6t/person/year

Rwanda (Howe and Gulick, 1980)
1976: Woodfuel consumption = .85t/person/year = 12.8 GJ/person/year
Woodfuel use = 96% of total energy use

Saudi Arabia (Pyle, 1980)
Biomass energy use = 0.1 GJ/person/year = less than 0.01t/person/year

Senegal (Howe and Gulick, 1980)
1976: Woodfuel use = .5t/person/year = 9 GJ/person/year
Woodfuel use was 63% of total energy use

(FRIDA, 1980)
Firewood consumption = 0.53 m^3 = .4t/person/year = 6GJ/person/year
Charcoal consumption = 163 kg/person/year = .16t/person/year = 4.8GJ

(US AID)
Charcoal consumption = 124 kg/person/year = .12t/person/year = 3.6 GJ

(Service Water and Forest)
Charcoal consumption = 251 kg/person/year = .25t/person/year = 7.5 GJ

Sierra Leone (Howe and Gulick, 1980)
1976: Woodfuel use = .8t/person/year = 12 GJ/person/year
Woodfuel = 76% of total energy use

(Richards, 1980)
Rough estimate - firewood use = .6 - .75t/person/year = 9 - 11.3 GJ
Does not include wood use for festivals, funerals, tobacco-curing or rice par-boiling

Somalia (Howe and Gulick, 1980)
1976: Woodfuel use = 1t/person/year = 15 GJ/person/year
Woodfuel = 90% of total energy use

(Pyle, 1980)
Biomass use is 3.6 times commercial energy use = 5 GJ/person/year = 0.33t/person/year

South Africa (Malan and Hattnigh, 1976)
In the Black Homelands in 1970, 7 million people consumed 5.4 million tons per year, or .8t/person/year

Sri Lanka (Frisch, 1980)
1975: Wood use = 3.3 million t wood
This was 54% of total energy use

Sudan (Arnold and Jongma, 1978)
Average wood use = 1.7 m^3/person/year = 1.2t/person/year = 18 GJ

(Vergara and Pimentel, 1978)
Fuelwood accounts for 65% of total energy use (estimated at an average of 10GJ/person/year)
Fuelwood accounts for 99.7% of energy produced in the country
Biomass - energy produced = 28 x 10^{12}kcal
= 1.6 x 10^6kcal/person
= 6.7 GJ/person
= 0.5t wood/person/year

(Khalifa, 1979)
Firewood consumption = 1.35 m^3/person/year plus
0.65 m^3 firewood equivalent charcoal, for a total of
2.0 m^3/person/year = 1.4 t/person/year = 21 GJ/person/year

(Howe and Gulick, 1980)
Woodfuel use = 1.3t/person/year = 20 GJ/person/year
Woodfuel use = 81% of total energy use

(Pyle, 1980)
Biomass use is 3.6 times commercial energy use = 15.2 GJ/person/year
= 1 t/person/year

Syria (Pyle, 1980)
Biomass energy use = 0.1 GJ/person/year = less than 0.01t/person/year

Tanzania (Openshaw, 1974)
Woodfuel consumption = 1.8t/person/year = 27 GJ/person/year
99% of population use woodfuel
92% of population is rural

(Openshaw, 1976, 1979)
Average woodfuel use = 2.13 m^3 (roundwood equivalent)
= 1.5t/person/year
= 22 GJ/person/year

(Arnold and Jongma, 1978)
Woodfuel use = 2.29 m^3 = 1.7t/person/year

(Tanzania National Scientific Research Council, 1979)
Fuelwood and charcoal use = 2 m^3/person/year = 1.4t/person/year
= 20 GJ/person/year

(Brown and Howe, 1978)
Biogas (single family plant) heat energy costs 1/9 of electricity

(FAO, 1976)
Fuelwood and charcoal use = 25.2 GJ/person/year = 1.7t/person/year

(Howe and Gulick, 1980)
Woodfuel use = 2.3t/person/year = 34 GJ/person/year
Woodfuel use = 94% total energy use

Thailand (Openshaw, 1976)
Fuelwood use = 1.1t/person/year = 16.5 GJ/person/year
97% of population use woodfuel

(Arnold and Jongma, 1978)
Woodfuel use = 1.36 m^3/person/year = 1 t /person/year
= 15 GJ/person/year

Thailand (Wibulswas, 1980)

Official estimation: wood and charcoal = 11% of total energy consumption. "But actual consumption figure may be much larger than the estimated figure".

Bagasse = 7% of total energy consumption

(Openshaw, 1972)

Bangkok (popn. = 4.5×10^6) used $3 \times 10^6 m^3$ wood
= $0.7\ m^3$/person/year = 0.5t wood/person/year
= 7 GJ/person/year

(FAO, 1972)

Rural areas: Woodfuel use = $1.33\ m^3$/person/year = 1t/person/year = 15 GJ
98% used woodfuel

Urban areas: Woodfuel use = $1.54\ m^3$/person/year = 1.1t/person/year = 17 GJ
87% used woodfuel

Countrywide: Woodfuel use = $1.36\ m^3$/person/year = 1t/person/year = 15 GJ
97% used woodfuel

Togo (Grut, 1971)

Total wood consumption = 763 kg/year including firewood equivalent of charcoal = .76t/person/year = 11.5 GJ/person/year

(FRIDA, 1980)

Firewood consumption = 946 kg/person/year = 14.3 GJ/person/year
Charcoal consumption = 160 kg/person/year = 4.8 GJ/person/year

Tunisia (Hamza, 1978)

Average wood use = $0.33\ m^3$/person/year
= 0.2t/person/year = 3 GJ/person/year

Rural areas (0.7×10^6 people) wood use = 1.2t/person/year
= 18 GJ/person/year

Urban areas (5×10^6 people) wood use = 0.1t/person/year

(Pyle, 1980)

Biomass use is 0.7 times commercial energy use = 9.4 GJ/person/year
= 0.63t/person/year

(Frisch, 1980)

1975: Wood use = 3.5 million tonnes.
This was 41.8% of total energy

Turkey (Cetelineli', 1977)

1972 - non-commercial energy was 30% of total energy use
1977 - non-commercial energy was 20% of total energy use

Upper Volta (Floor, 1977)

Woodfuel is 94% of total energy consumption
Per capita fuelwood consumption = $0.7\ m^3$/year = 0.5t/person/year
Estimated fuelwood consumption (1970) is $3 \times 10^6 m^3$ for rural areas (90% of the population)
and $3 \times 10^5 m^3$ for urban areas (10% of the population)
Spend 20% of wages on fuelwood or charcoal

(Bertrand, 1975)

Firewood consumption = 675 kg/person/year = .68t/person/year = 10.2 GJ
Charcoal consumption = 5 kg/person/year = .2 GJ/person/year

(FRIDA, 1980)

Firewood consumption = $0.78\ m^3$/person/year = 546 kg/person/year = 8.3 GJ
Charcoal consumption = 147 kg/person/year = 4.5 GJ/person/year

Upper Volta (Cont'd)

(Langerhorst et al, 1977)
Firewood consumption = 600 kg/person/year = .6t/person/year = 9 GJ

(Ouedraogo, 1974)
Firewood consumption = 475 kg/person/year = .48t/person/year = 7.2 GJ

(Goudet, 1964)
Firewood consumption = 550 kg/person/year = .55t/person/year = 8.3 GJ

Yemen Arab Republic (Pyle, 1980)
Biomass use is 4.2 times commercial energy use = 5.1 GJ/person/year
= 0.33t/person/year

Yemen People's Democratic Republic (Pyle, 1980)
Biomass use is .55 times commercial energy use = 5.0 GJ/person/year
= 0.33t/person/year

Venezuela (Frisch, 1980)
1975: Wood use = .28 million t.
This was .4% of total energy use.

Zambia (Mabonga Mwisaka, 1978)
Charcoal use = 1.25×10^5t (= 1.5×10^6t wood)
= 4.75×10^{15}J = 0.9 GJ/person/year
Woodfuel use = 2.3×10^6t = 0.45t/person/year = 6.8 GJ/person/year
Combined wood use = 3.8×10^6t = 0.8t/person/year
35% of total energy use is firewood and charcoal.
There are 632,000 rural families of about five persons each-
this equals 3.16 million people or 62% of the population.
Each family uses about 10 kg firewood/day or 3.65t/year
(= .73t/person/year in rural areas)
Urban areas - 60% (1.13 million people) depend on charcoal for household use.
Charcoal and coal have same energy costs; electricity costs 5x as much and kerosene costs 16x as much.
Copperbelt Province - 2200 people engaged in charcoal industry,
producing 52,000 t charcoal per year. Copper refineries use
100t charcoal per day.

(Frisch, 1980)
1975: Wood use = .8 million t. This was 9.4% of total energy use.

Zimbabwe (Banks,1980)
81% total wood use (= 6 million m^3) is fuelwood
57% total consumption (= 3.5 million m^3) is used in Tribal Trust Lands
Most fuelwood currently used is obtained from land being cleared for agriculture.

(Johnston, 1980)
Woodfuel = 28% of total energy consumption
Bagasse = 2% of total energy consumption
Biomass thus accounts for 30% of total energy consumption.
Woodfuel consumption = 0.83 m^3/person/year = .6t/person/year = 9 GJ

(Whitlow, 1980)
6 million m^3 of wood are consumed each year.
0.6 million m^3 timber used in mining industry, mostly as fuelwood
for mine-workers. 60 - 70 m^3 wood are trucked into Salisbury yearly,
from sources up to 200 km away. It may require 20% of household earnings
to purchase woodfuel to cook and provide heat.

Recent information received from

UN Conference on New and Renewable Sources of Energy

(UNERG : Nairobi, 10-21 August, 1981)

Australia (A/CONF/NR/42)

3% of primary energy requirements come from biomass: 1% from firewood
2% from bagasse

Bangladesh (A/CONF/NR/68)

About ⅔ of total energy needs are met by traditional fuels.
Import of petroleum (18% of energy needs) claimed about ⅔ of total export earnings for 1980/81.

Benin (A/CONF/NR/32)

Wood use = 150,000 m t/year
25 MW hydropower

Brazil (A/CONF/NR/55)

1979 data: Ethanol provided 1.6% of total energy consumption
Firewood provided 17.4% of total energy consumption
Charcoal provided 2.6% of total energy consumption
Sugar cane bagasse provided 4.7% of total energy consumption
Hydropower provided 28.3% of total energy consumption
A total of 54.6% of energy was from renewable sources

Canada (A/CONF/NR/62)

Biomass = 3% of total energy consumption
About 50% of wood mill wastes generated are used as fuel.
Canada has one of the world's largest peat reserves, but development has not been pursued yet.
Hydropower = 60% of electricity generation, or 24% of total energy consumption.

China (A/CONF/NR/22)

Biomass provides more than 25% of energy consumption and equals 220 mtoe.
Fuelwood and straw are the main sources.
1978: There were 6.39 m biogas digesters, mainly in Sichuan. More than 40 institutes conduct research on biogas.
100,000 m^2 of solar collecters are in use
China has 8.5% of the world's peat resources

Cuba (A/CONF/NR/26)

Sugarcane bagasse = 25% of total energy
100,000 m^3 alcohol from sugarcane is produced annually

Dominican Republic (A/CONF/NR/6)

Petroleum = 67.1% of total energy
Residues = 1.3% of total energy
Firewood = 11.6% of total energy
Bagasse = 19.2% of total energy
Hydropower = 0.8% of total energy
Preparation of meals: City dwellers use 83% of total country-wide propane gas consumption for this purpose, while rural inhabitants use 96% of firewood used.

Egypt (A/CONF/NR/41)

Non-commercial energy = 15% of total energy consumption. In 1975, one-fifth of total energy consumption was provided by non-commercial sources such as dung cakes and agricultural wastes. Consumption of petroleum products is increasing very rapidly, almost doubling every six years - petroleum products subsidised.

Finland (A/CONF/NR/3)

In 1960, wood,hydropower accounted for 60% of total primary energy production. These now account for about 30%
Peat = 2% of total energy consumption
Wood = 16% of total energy consumption
Hydropower = 11% of total energy consumption
Peat is Finland's largest energy reserve, equal to 5000 mtoe
30% of energy research ($17 m in 1980) went towards wood and peat

France (A/CONF/NR/7)

Renewable energy = 9% of total energy in 1979 (about 80% hydro and 20% wood)
In 1980 wood provided 3 Mtoe (=8 4 mt wood)

Guyana (A/CONF/NR/44)

Bagasse, wood, rice chaff and charcoal = 15% of national energy consumption
Bagasse = 450,000 t available in 1979
Rice chaff = 12,000 t available in 1979
Firewood = 38,000 t available in 1979
Charcoal = 1,000 t available in 1979 (Produced by pit tumulus method)

India (A/CONF/NR/)

Non-commercial energy = 40% of total energy today
Total consumption of firewood in 1975-76 = 133 m t
Total consumption of dung cakes = 73 mt in 1975-76, mostly in household sector
There are approximately 80,000 biogas plants in India

Indonesia (A/CONF/NR/52)

Estimate that firewood consumption equals about 50% of total energy use and 80% of energy use in rural areas. Firewood plantations were started in 1975-some on suitable soils produce 70 - 120 m^2/ha/yr, although some retarded growth has been observed - the cause of this is under investigation.
The target is to establish 300,000 ha of new forest per year and to "green" 70,000 ha of critical private land (to combat erosion and floods).
Success rate - about 50% for reforestation and 30% for greening. There is free distribution of 15-20 firewood species.
Up to 1980 - 65,810 ha of land had been planted with fast growing trees.

Ireland (A/CONF/NR/28)

16% of national energy requirements are met by peat.

Israel (A/CONF/NR/19)

Solar energy provides 1.2% of total energy consumption

Jamaica (A/CONF/NR/16)

Imported petroleum = 99% of commercial energy requirements
50% of this is used by bauxite and alumina industry (which produce ⅔ of the country's export earnings).

Japan (A/CONF/NR/49)

Firewood and charcoal = .1% of total energy in 1979
Hydropower = 5.1% of total energy

Jordon (A/CONF/NR/61)

Wood and dung = .5% of total energy

Kenya (A/CONF/NR/36)

1979: Fuelwood and charcoal = 75% of total energy consumption
(30m m^3 = 7mtoe)
Hydropower = 4.3% of total energy consumption

Liberia (A/CONF/NR/8)

Oil = 55.4% of total energy consumption
Hydropower = 12.2% of total energy consumption
Firewood = 21.2% of total energy consumption
Charcoal = 11.3% of total energy consumption
A 1979 survey of selected cities and towns put the cost of firewood, charcoal and kerosene for a family of five at $262 per year. As GDP per capita is $514.8, about 10% of the family budget is spent on wood, charcoal and kerosene. Of the three, charcoal is the preferred commodity.
In Monrovia, av. per capita use of wood is 0.04 m^3/year, of which 47.9% is in the form of charcoal. In rural towns such as Samyea, up to 5.5 m^3 wood/year is used per capita (only 0.15% is charcoal)

Norway (A/CONF/NR18)

Hydropower provides 56% of energy consumed
Wood provides 0.6% of energy consumed

Peru (A/CONF/NR/17)

Wood, charcoal and vegetable wastes provide 30.7% of total energy consumption (oil = 50.6%, natural gas = 9.9%, hydroelectric = 6.3%, coal = 2.8%, wood and charcoal = 24.7%, vegetable wastes = 5.7%)

Sierra Leone (A/CONF/NR/38)

1980: Crude oil = 20% of national income or 30% of foreign exchange budget
Firewood consumption = 2.5 m m^3/year of which 250,000 is supplied to Freetown

Sudan (A/CONF/NR/11)

1979-80: Wood = 69.1% of total energy
Charcoal = 6% of total energy
Crop and animal wastes = 1.4% of total energy
In 1979-80, 10.65 mt of firewood was used. 98% was used as household fuel, 2% for small industry such as brick burning, pottery, tobacco curing and production of steam.
Estimates it needs to plant 120,000 ha of trees but only 7000 ha (6% of target) are being planted. Wood deficit projected for 1985.

Suriname (A/CONF/NR/69)

17,000 t ricehusks used per year for electricity production(17000 barrels oil saved)
Sugarcane estate burns 24000 t bagasse per year (50,000 barrels oil saved)
Oil palm fibre and waste- 3000 t burnt (replacing 6000t oil)(increasing)
15,000 t wood wastes burnt- saving 30,000 barrels oil
Total crops residue and wood waste replaces some 103,000 barrels oil/year.
Hydropower = 20.6% of total energy
Biomass = 1.5% of total energy
73.5 % of total energy is used by bauxite industry

Sweden (A/CONF/NR/21)

1979: Hydropower = 27% of total energy
Wood = 6.5 % of total energy
Peat resources = 3000 Mtoe
1979: 4 m m^3 of wood and wood chips were used as fuel = 0.7 Mtoe

Thailand (A/CONF/NR/65)

1979 demand for firewood was 30-40 m m^3.
Bagasse = 8.2% of total energy in 1979 (mostly used in sugar industry)
Rice husk> 1% of total energy in 1979 (mostly used in small industry and for cooking)

<u>Turkey</u> (A/CONF/NR/63)
1980: Firewood consumption = 14.10^6 tons constituting 12.5% total energy consumption

<u>U.S.S.R.</u> (A/CONF/NR/51)
Peat reserves = 157200 m t. 1980 output = 84 m t (of which 35 mt used in agriculture, 27 mt used in energy generation and 9.2 mt used to produce peat briquettes)
Biomass = > 1% total energy consumption

<u>Zambia</u> (A/CONF/NR/64)
Wood - energy source for 90% of population(both urban and rural)
500,000t charcoal was used in 1978 (equivalent to 5.0 m t raw wood)
Wood use in 1978 = 464,000 t
Improbable that biogas would be socially acceptable- one trial digester.

<u>Nigeria</u> (A/CONF/NR/59)
Firewood = more than 40 % of total energy, mostly for domestic cooking.
Estimate of firewood use = 70 m m^3 (minimum figure)
Fuelwood plantations have continued to be established since 1912. However, severe pressure on land has reduced these.

References for Energy Data Tables

Afolabi Ojo, G.J.(1979) Some Socio-Economic Determinants of the Use of Forest Resources for Energy and Organics: A Nigerian Situational Overview. Pp. 175-178 in UNESCO, Proceedings of an International Workshop: Biological & Sociological Basis for a Rational Use of Forest Resources for Energy & Organics. Published by U.S. Dept. of Agriculture, S.E. Experimental Station, Asheville, N.C.

Argal,P.(1978) Role of Wood Energy in the Rural Economy of India, Eighth World Forestry Congress, Jakarta, 16-28 October, 1978. 14 pp.

Arnold & Jongma (1978) Fuelwood and Charcoal in Developing Countries, Unasylva, Vol. 29, No.118, pp 2-9.

Ay, P. (1978) Fuelwood and Charcoal in the West African Forest: Field Research in WesternNigeria in: Morgan, Moss, Ojo (1978) Rural Energy Systems in the Humid Tropics, UN University, Rural Energy Systems Project, Ife, Nigeria, 10-12 August, 1978.

Banks, P.F.(1980) The Indiginous Woodland - A Diminishing Source of Fuelwood, Energy Symposium '80, Salisbury, Zimbabwe, October 1980. 7pp.

Bertrand, A.(1977) Les problems du bois de chauffage et du charbon du bois en Afrique Tropicale, Bois et Forets des Tropiques, 1977, No. 173, pp 38-48

Bertrand, A.(1975) Besions en bois de la ville de Ouagadougou - Amenagement des vallees de Volta, Centre Techniques Forestier Tropical, Jan. 1975.

Bertrand, A.(1979) Consommation et utilisation des combustibles ligneux en Afrique Occidentale et Centrale, Centre Technique Forestier Tropical, Report for FAO, Rome, April 1979.

Best,M.G.(1979) The consumption of energy for domestic use in three African villages, MSc. thesis, University of Cape Town, South Africa. 144 pp.

Bialy,J.(1979) Firewood Use in a Sri Lankan Village, A Preliminary Study, School of Engineering, University of Edinburgh, U.K. April 1979. 37 pp.

Brattle,R.V.(1979) Energy and Cooking in the Sudan - A Summary of a Literature Survey and Field Experience, University of Surrey, Guildford, U.K.

Briscoe,J.(1979) Energy Use and Social Structure in a Bangladesh Village, Population and Development Review, Vol. 5, No. 4, pp 615-641.

Brown and Howe (1978) Solar Energy for Village Development, Science, Vol. 199, 10 February 1978, pp 651-657.

Castan,R. (1966) Amenagement forestier pour le revitaillement des populations en bois de feu et en bois de service, Centre Technique Forestier Tropical

Caude,G.(1977) Les besoins energetiques en milieu rural Malien,SEMA (France)

Ceteiincelik,M.(1977) in Energy International, Vol.14, No. 9, September 1977, pp 27-30. Reprinted in Smil & Knowland, which see.

Cline-Cole,R.E.A.(1979) Firewood Demand and Supply, Waterloo, University of Sierra Leone, dissertation for BA, Degue, 1979. 163 pp.

Commonwealth Science Council(1979) Report on the Seminar on New Energy Technologies, 12-13 September, Georgetown, Guyana. CSC, Marlborough House, Pall Mall, London SW1X 5HX

Cox,W.(1978) Energy Consumption and Economic Growth: A Study of the Barbadian Experience: 1960-1970, Central Bank of Barbados, Research Department

Delwaulle and Roederer (1973) Le bois de feu a Niamy, Bois et Forets des Tropiques, Nov.-Dec. 1973, No. 152, pp 55-60.

Digernes,T.(1979) Report from African Solar Energy Workshop, 21-26 May, 1979, Atlanta, Georgia. DERAP Working Papers, The Chr. Michelsen Institute, Fantoftvegen 38-N 5036 Fantoft-Bergen, Norway.

Digernes,T.(1979) Fuelwood Crisis Causing Unfortunate Land Use - And the Other Way Around, Norsk. Geogr. Todsslr., Vol. 33, pp23-32 Oslo

Donovan, D.G.(1981) Fuelwood: How Much Do We Need? DGD-14, Institute of Current World Affairs, 4 West Wheelock St, Hanover, N.H. 03755 USA

Earl, D.E.(1975) Forest Energy and Economic Development, Clarenden Press, Oxford, U.K.

Ernst, E.(1978) Fuel Consumption Among Rural Families in Upper Volta, West Africa, Eighth World Forestry Congress, Jakarta, 16-28 Oct. 1978

FAO/World Bank (1979) Draft of the Malawi Wood Energy Project, FAO/World Bank Cooperative Programme, Investment Centre, FAO, Rome

FAO (1978) Yearbook of Forest Products, 1965-1976, FAO Forestry Series, FAO Statistics Series, FAO, Rome

Floor,W.M.(1977) The Energy Sector of the Sahelian Countries, Policy Planning Section, Ministry of Foreign Affairs, The Netherlands

FRIDA Investments (1980) Domestic Energy in Sub-Saharan Africa, Dec. 1980. Grosvenor House, 141-143 Drury Lane, London,WC2B 5TD

Frisch, J.R. (1980) Survey of the Long-Range Development of the Energy Supply and Demand of Less Developed Countries, 11th World Energy Conference, 34 St. James St, London, SW2A 1HD

Gambia Land Resources Project (1973) The Gambia: A Wood Consumption Survey and Timber Trend Study: 1973-2000. K.Openshaw's unpub. report to ORD/LRD Gambia Land Resources Development Project, Midlothian, U.K. Cited in Smil & Knowland, which see.

Gatera,F.(1978) Le bois et le charbon de bois dans le commune urbaine de Nyarugenge, Kigali, Mem. de Licence,178. 111pp.

Ghosh,R.C.(1979) The Socio-Economic Consequences and Constraints to the Land and Forest for Energy and Organics in India,Biological and Sociological Basis for a Rational Use of Forest Resources for Energy & Organics, Michigan State University and UNESCO, Pub. by U.S. Dept. of Agriculture, S.E. Experimental Station, Asheville, N.C. May 1979.

Gibert, G.(1978) Le revitaillement de Brazzaville en bois de chauffage, Bois et Forets des Tropiques, 1978, pp 19-36

Goudet,J.P.(1964) Produits ligneux - produits de cueillette ressources - satisfactions de besoins dans le perimetre de Ouahigouya, Centre Technique Forestier Tropical Section Niger - Haute Volta, August 1964

Grut,M.(1972) Market for firewood, poles and sawnwood in the major towns and cities in the Savannah Region, FAO, Rome, Technical Report 6: (FO:SF/NIR/16) 1972, 46 pp

Grut,M.(1971) Marche du charbon de bois, du bois de chauffage ét du bois rond au Togo, PNUD/FAO (FO:SF/TOG 10) Technical Report 6, Rome 1971

Hamza,H.(1978) Biological and Sociological Use Basis for National Use of Forest Resources for Energy and Organics in Arid and Semi-Arid Areas in Tunisia, North Africa, pp 180-2 in: UNESCO, Proceedings of an International Workshop: Biological & Sociological Basis for a Rational Use of Forest Resources for Energy and Organics, Michigan St. University, May 1979, Pub. by U.S. Dept. of Agriculture, S.E.Experimental Station, Asheville, N.C.

Howe and Gulick(1980) Fuelwood and other renewable energies in Africa, a progress report on the problem and response, L'energie dans les communautes rurales des pays du Tiers Monde, C.N.R.S. Domaine Universitaire de Bordeaux 33405 Talence, France. 24 pp

Hughart,D.(1979) Prospects for Traditional and Non-Conventional Energy Sources in Developing Countries, World Bank Working Paper No.346. July 1979. 137 pp. World Bank, 1818 H St, NW, Washington, D.C. USA

Ibrahim,S.M.A.(1979) Energy in Egypt: Resources, Uses and Outlook to 2000, Energy, Vol. 4, pp 1069-1077.

Jackson,J.(1980) Wood Fuel in Malawi, Energy Symposium '80, Salisbury, Zimbabwe, Oct. 1980. 12pp

Johnston, J.C.(1980) Wood Fuel - A Neglected Energy Source in Zimbabwe, Energy Symposium '80, Salisbury, October 1980. 9 pp

Joshy,D.(1979) Letter of 16/9/79 concerning domestic energy consumption in Nepal, Div. of Soil Science & Agricultural Chemistry, Khumaltar, Lalitpur, Kathmandu, Nepal.

Kabagambe,D.M.(1976) Aspects of Resource Conservation - the Role of the Charcoal Industry in Kenya, Institute for Development Studies, Working Paper No. 271, University of Nairobi, 1976.

Khalifa,K.O.(1979) Energy from Sudan Forests- Wood for the People - SCC Consultation on Development, March 1979. Sudan Council of Churches

Kokwaro,J.O.(1979) Indigenous and Introduced Common Firewood and Charcoal Plants of Kenya, Energy and Environment in East Africa, 7-11 May 1979, The Royal Swedish Academy of Sciences, Stockholm

Langerhorst,Prast, Thalkhammer and Floor(1977) Solar Energy Report on a Study of the Difficulties Involved in Applying Solar Energy in Developing Countries, Ministry for Development Cooperation, The Hague

Mabonga-Mwisaka,J.(1978) Wood Energy and Rural Communication in Zambia, Eighth World Forestry Congress, Jakarta, October 1978

Makijani and Poole(1975) Energy and Agriculture in the Third World, Ballinger Publishing Co, Cambridge, Mass. 168 pp

Malan and Hattingh(1976) Black Homelands in South Africa, African Institute of South Africa, Pretoria

Marguand and Githinji(1979) Energy Resources in Kenya and their Environmental Impacts, International Seminar on Energy and Environment in East Africa, 7-10 May 1979 Nairobi (authors c/o Dept. of Mechanical Engineering, University of Nairobi)

McComb and Jackson(1970) Role of Tree Plantations in Savannah Development - Technical and Economic Aspects, 2nd Session, FAO Committee for the Developing Tropics, 1970. FAO, Rome

Ministry of Energy(1979) White Paper - Energy Policy and Planning for Papua New Guinea, Ministry of Minerals and Energy,P.O.Box 2352, Konedobu

Muchiri,G.(1978) Rural Energy Needs and Alternative Sources, Biogas Seminar, Egerton College, 5-8 Sept. 1978(author c/o Dept.of Agricultural Engineering, University of Nairobi)

Mukhtar, (1978) Woodfuel as a Source of Energy in Sudan, Forests Administration, Khartoum. First Energy Conference, 15-17 April 1978. 19 pp

Murrugappa Chetiar Research Centre & University of Strathclyde,(1981) Self-Reliant Development - an IFIAS Sponsered Study, Final Status Report, March 1981. 144 pp

Muthoo, M.K.(1978) Forest Energy and the Brazilian Socio-Economy with Special Reference to Fuelwood, Eighth World Forestry Congress, Jakarta, Oct. 1978. 14 pp(M.K.Muthoo c/o UNDP. C.P. 07-0285, Brasilia, Brazil)

Mutula Mungala,P.(1979) Estimation of Present and Likely Future Demand for Fuelwood and Charcoal in Machakos District, Kenya. Thesis, University of Dar-Es-Salaam, Kenya

Nyoike, (1979) Kenya's Energy Situation at Present and in the Future, Royal Institution Forum "Third World Energy Strategies and the Role of Industrialized Countries"June 1979. Royal Institution, 21 Albemarle Street, London W1X 4BS U.K.

Openshaw,K.(1974) Woodfuels for the Developing World, New Scientist, 31 January 1974, pp 271-272

Openshaw,K.(1976) Woodfuel - A Time for Re-Assessment pp 110-126 in Khamala and Castelino,eds, Energy Resources in East Africa, Proc. of the 12th Annual Symposium of the East African Academy, August 1976 Kenya National Academy for Advancement of Arts and Sciences, Nairobi

Openshaw and Moris(1979) The Socio-Economics of Agroforestry in Chandler and Spurgeon, eds, Conference on International Cooperation in Agroforestry, ICRAF, P.O.Box 30677 Nairobi, Kenya

Ouedraogo,M.M.(1974)L'approvisionnement de Ouagadougou en prodruits vivriers, en eau et en bois, Universite de Bordeaux III et Centre Voltaique de la Recherche Scientifique. These de Doctorat - IIIme cycle, 1974

Pare,J.(1979) Unite Ecologique Experimentale de Maradi: Essai de Synthese, GERDAT, 1979, quoted in Raynaut,C.(1980) Collect du Combustible et Equilibre des Relations avec le Milieau Natural dans les Communautes Villegoises du Departement de Maradi, in: L'energie dans les communautes rurales des paus du Tiers Monde, C.N.R.S. Domaine Universitaire de Bordeaux 33405 Talence, France

Pyle,D.L.(1980) Biomass Energy in the Arab World, Dept. of Chemical Engineering, Imperial College, London

Ravindranath et al(1980) An Indian Village Agricultural Ecosyatem, Case Study of Ungra Village parts I & II, ASTRA, Indian Institute of Science, Bangalore 560 012 India

Reddy,A.K.(1979) Alternative and Traditional Energy Sources for the Rural Areas of the Third World, Royal Institution Forum "Third World Energy Strategies and the Role of Industrialized Countries, Royal Institution, 21 Albemarle St, London W1X 4BS U.K.

Revelle,R.(1976) Energy Use in Rural India, Science, Vol. 192, pp 969-975

Richards,P.(1980) Farming System Energy Requirements - Implications for African Rural Development, Bordeaux 1980- L'energie dans les communautes rurales des pays du Tiers Monde, C.N.R.S. Domaine Universitaire de Bordeaux 33405 Talence, France

Sankare,O.(1979) Analyse de la consommation et de besoins en combustibles ligneux du district de Bamako, Rapport de Synthese, Institute Polytechnique Rural de Katibougou, Mali for FAO, Rome

SIECA(1978) Energy Planning Publication, March 1978, plus CEPAL(Comision Economica para America Latina) data, quoted in Gewald,(1980) The Importance of Fuelwood in Central America, L'energie dans les communautes rurales des pays du Tiers Monde, C.N.R.S. Domaine Universitaire de Bordeaux 33405 Talence, France.

Smil,V.(1976) China's Energy: Acheivements,Problems, Prospects, Praeger Special Studies in International Economics and Development, Praeger Publishers, N.Y. 219 pp

Smil and Knowland, eds,(1980) Energy in the Developing World: The Real Energy Crisis, Oxford University Press,Oxford, U.K. 386 pp

Spurgeon,D.(1979)Full Steam Ahead to Save Kenya's Fuel Nature,Vol.281, 11 October 1979 page 423.

Suarez,C.E.(1979) Energy Problems in the Developing Countries and Their Repercussion on the Industrialized Countries, Royal Institution Forum Third World Energy Strategies and the Role of the Industrialized Countries, Royal Institution, 21 Albemarle St,London W1X 4BS U.K.

Tall,S.(1974) L'economie du charbon du bois a Dakar, Dakar, Faculty Lettres et Sciences Humaines, T.E.R. 1974 89pp

Tanzania National Scientific Research Council(1979) Tanzania Energy Profile and Development of Alternative Energy Systems, African Energy Group, Arusha, Tanzania, September 1979

Thipe and Mokobi(1979) Current Status of Energy Needs and Field Experiences with Solar Techniques in Botswana, African Energy Group Meeting, Sept. 1979 Arusha Tanzania (author c/o Ministry of Mineral Resources and Water Affairs, Private Bag 0018 Gaborone Botswana)

Uhart,E.(1976) The Wood Charcoal Industry in Africa, African Forestry Commission, March 1979, FAO

Vennetier,P.(1979) La comsommation d'energie Traditionelle en milieu Africain. L'exemple de Ngaoundere, Centre d'Etudes de Geographie, Talence, Cedex, France

Vennetier,P.(1980) Les Problems du Bois de Feu et du Charbon de Bois en Afrique Tropical, Analyse et Bilan des Recherches, Colloque: L'energie dans les communautes rurales des pays du Tiers Monde, May 1980 C.N.R.S Domaine Universitaire de Bordeaux 33405 Talence, France 20 pp

Vergara and Pimentel (1978) Fuels from Biomass: Comparative Study of the Potential in Five Countries: The United States, Brazil, Sudan, India and Sweden, Advances in Energy Systems and Technology, Vol.1, pp125-173

Vojdani and Woolam (1978) Possibilitites for Solar Energy Applications for Rural Iran, Materials and Energy Research Centre, Tehran,P.O.Box 41-2927 Tehran, Iran.

Western and Ssemakula (1979) The Present and Future Patterns of Consumption and Production of Wood Energy in Kenya, International Workshop on Energy and Development, Nairobi, May 1979

White,R.(1979) A handbook for village woodlot planning and management in Botswana, Report for Village Areas Development Project, Government Printer, Gaborone

Whitlow,J.R.(1980) Deforestation in Zimbabwe - Supplement to Zambezia, 1980, Journal of the University of Zimbabwe, Salisbury

Wibulswas,P.(1980) Energy for Rural Communities in Thailand, Colloque: L'energie dans les communautes rurales des pays du Tiers Monde, C.N.R.S. Domaine Universitaire de Bordeaux 33405 Talence, France Quotes from "Draft Plan on the Development of the Alternative Sources of Energy"(in Thai), Thai Nat'l Energy Administration, Bangkok, 1980

Winterhalder, Larsen and Thomas (1974) Dung as an Essential Resource in a Highland Peru Community, Human Ecology, Vol. 2, No. 2, 1974, pp 89-104

Appendix B

Institutions Involved in Alternative Energy Research

ALGERIA

Organisme National de la
Recherche Scientifique,
Station de l'Energie Solaire,
Route de l'Observatoire,
Bouzareah, Alger

ARGENTINA

Centro de Estudios Fotosinteticos
y Bioquimicos (CEFOBI)
Consejo Nacional de Investiga-
ciones Cientificas y Technicas.
Fundacion Miguel Lillo-Universidad
Nacional de Rosario,
Suipacha 531, 2000 Rosario

Instituto de Economia Energetica(IDEE)
Casilla de Correo 138
8400 Bariloche

Grupo de Energia Solar (GES)
Universidad Nacional,
4400 Salta,
Provincia de Buenos Aires

Solar Energy Division,
Comision Nacional de Energia Atomica,
Avd. del Libertador 8250
1429 Buenos Aires.

AUSTRALIA

Appropriate Technology and
Community Environment (APACE)
P.O.Box 770, North Sydney,
New South Wales 2060

Commonwealth Scientific and
Industrial Research Organization(CSIRO)
Box 225, Dickenson, A.C.T.2602

Development Studies Centre
The Research School of Pacific Studies,
Australian National University,
P.O.Box 4, Canberra, A.C.T. 2600

Energy Research Centre,
University of Sydney, Sydney,
N.S.W. 2006

AUSTRALIA (cont.)

CSIRO Division of Building Research,
P.O.Box 56,
Highett, Victoria 3190

Appropriate Technology Development
Group, 20 Holdsworth Street,
Woolahara,
New South Wales 2025

AUSTRIA

Institute for Environmental Research,
Hilmteichstrasse 75A
A-8010 Graz

United Nations Industrial
Development Organization (UNIDO)
Lerchen Felder Strasse 1,
P.O.Box 707,
A-1070 Vienna

BANGLADESH

Chemical Engineering Department
Bangladesh University of Science
and Technology
Dacca 2

BARBADOS

National Council for Science
and Technology,
Flodden Culloden Road,
Bridgetown

University of the West Indies,
Biology Department
Cave Hill Campus,
Bridgetown

BELGIUM

Appropriate Technology for
Developing Countries (ATOL)
Blijde Irkomstraat 9, 3000
Leuven

Association Internationale de
Developpement Rural,
20 rue du Commerce, Boite 9,
B-1040 Brussels

BELGIUM (cont.)

Universite Catholique de Louvain
Faculte des Sciences Appliquees
Batiment Stevin
Place du Levantz,
1348 Louvain-La-Neuve

BOLIVIA

Instituto de Energie
Academy of Sciences,
Casilla 5279, La Paz

BOTSWANA

Rural Industries Innovation Centre,
Box 138, Kanye

Village Industries Program,
P.O.Box 464, Gaborone

National Institute of Development and
Cultural Research, University of
Botswana and Swaziland,
Private Bag 0022, Gaborone

BRAZIL

Centro de Technologia Promon,
Praia do Flamengo 154-12 andar,
20 000 Rio de Janeiro

Laboratory of Solar Energy,
University Campus,
Paraiba Federal University,
Joao Pessoa, 58000 Paraiba

Ministerio das Minas e Energia
Esplanada dos Ministerios - Bloco J
70056 Brasilia (DF)

PPGEMM-Escola de Engenharia,
Universidade Federal do Rio
Grande do Sul,
Av. Osvaldo Aranha, 9º-6 andar,
90 000 Porto Allegre, R.S.

Technology Center of Minas Gerais,
Avenida Jose Candido da Silveira 2000,
Horto 30 000, P.O.Box 2306,
Belo Horizonte, Minas Gerais

BRAZIL (cont.)

Universidade Estadual de Campinas
Instituto de Biologia, Depto de
Fisiologia Vegetal, Caixa Postal
1170, 13,100 Campinas, S.P.

Universidade Federal do Ceara
Nucleo de Fondes Nao Convencionais
de Energia, Centro de Ciencias,
Campus do Puci, Caixa Postal 935
60 000 Fortaleza-CE

BULGARIA

Scientific Production Economic
Combine "New Energy Sources"
Studentoki Grad, Bloc 19,
Sofia 1100

Central Laboratory of Solar Energy
and New Energy Sources,
Bulgarian Academy of Sciences,
72 Bd. Lenin, Sofia

New Energy Sources and Scientific
Production Institute,
Department of Science,
1100 Sofia

BURUNDI

Departement de l'Energie,
Gouvernement du Burundi,
B.P. 1860, Bujumbura

Groupe de Recherche des Utilisation des Energies Alternatives,
(GRUEA) Universite de Burundi,
Faculte des Sciences, B.P.2700
Bujombura

CAMEROON

Unite de Recherches Energetiques
de l'Onarest,
B.P. 4110, Yaounde

Centre for Energy Research,
National Office for Scientific
and Technical Research,
Yaounde

CANADA

Brace Research Institute
Macdonald College of McGill University,
Ste. Anne de Bellevue,
Quebec, H9X 1C0

International Development Research
Centre (IDRC)
Box 8500, Ottawa, K1G 3H9

Biomass Research and Development,
National Research Council,
Bldg. M-50, Montreal Road,
Ottawa ON K1A OR6

Conservation and Renewable Energy,
Ministry of Energy,
56 Wellesley Street,
Toronto, Ontario M7A 2B7

Conservation and Renewable Energy
Branch, Dept. of Energy, Mines and
Resources, 580 Booth Street,
Ottawa K1A OE4

Energy Research Group,
Carleton University,
C.J.Mackenzie Building, Rm. 218,
Colonel By Drive, Ottawa K1S 5B6

Institute of Man and Resources
50 Water Street,
P.O.Box 2008,
Charlottetown, Prince Edward Island

National Resources Institute
University of Manitoba,
Winnipeg, Manitoba R3T 2N2

Ryerson Energy Centre,
Ryerson Polytechnic Institute,
50 Gould Street,
Toronto,Ontario M5B 1E8

CAPE VERDE

Conservation and Development of
Natural Resources,
Ministry of Rural Development.

CHILE

Comision Economica para America
Latina, Casilla 179D
Santiago

Comision Nacional de Energia
Teatinos 120, Piso 14,
Santiago Correo 21, Casilla 14

Solar Energy Program
Institute of Technological Research,
P.O.Box 667, Santiago

PEOPLES REPUBLIC OF CHINA

The Beijing Institute of New
Technology Applications,
Shi Zhi Men, Beijing

Guangzhou Institute of Energy
Conversion,
Chinese Academy of Science,
81 Martyr's Road, Guang Zhou,
Canton

Shanghai Institute of Fossil Fuel
and Solar Energy Utilization,
1960 Longhua Lu, Shanghai, 200032

Zhejiang Provincial Solar Energy
and Biomass Institute,
Hangchow, Zhejiang.

COLOMBIA

Institute for the Development of
Renewable Natural Resources,
Carrera 14, No. 25 A-66,
Bogota, D.E.

Solar Energy Group
Dept. of Physics,
Universidad Nacional de Columbia,
Apartado Aereo 100100, Bogota 10.

DOMINICAN REPUBLIC

Santo Domingo Autonomous University
Calle Primera No.5
El Cacique III
Santo Domingo

COSTA RICA

Natural Renewable Resources Programme,
Centro Agronomico Tropical de
Investigacion y Esenanza,
Apartado 74, Turrialba

Instituto Tecnologico de Costa Rica,
Centro de Informacion Technologica,
Apartado 159, Cartago

ECUADOR

Instituto Nacional de Energia
Stam Prisca 233, Quito

Latin American Energy Organization
(OLADE) Casilla 119-A, Quito

EGYPT

National Research Centre,
Solar Energy Laboratory,
El Tahrir Str, Dokki, Cairo

Physics Dept., Solar Laboratory,
Faculty of Science,
Cairo University, Giza, Cairo

University of Alexandria
Faculty of Engineering - Solar
Application Laboratory
El-Hadara, Alexandria

ETHIOPIA

Ethiopian National Energy Commission,
P.O. Box 486,
Addis-Ababa

Ethiopian National Energy Committee,
P.O. Box 6690,
Addis-Ababa

Natural Resources Research Sub-
Council,
Ethiopian Science & Technology
Commission, P.O. Box 2490,
Addis-Ababa

FIJI

Institute of National Resources
University of the South Pacific,
P.O. Box 1168, Suva

FINLAND

Planning and Marketing Office,
Technical Research Centre of
Finland, Vuorimiehentie 5,
02150 Espoo 15

FRANCE

Applications of Research on
Energy and Society (ARES)
39, Rue Croix-Baragnon,
31000 Toulouse

Centre International de Recherche
sur l'Environnement et de le
Developpement (CIRED)
54, Boulevard Raspail,
75270 Paris Cedex 06

Comite Francais des Concours
d'Inventions et Innovations
Adaptees aux Regions en
Developpement (CIARD)
42, Rue Cambronne, 75015 Paris

Commissariat a l'Energie Solaire
208 Rue Raymond Losserand
75014 Paris

Groupe Energie et Biomasse
5 Rue Ste. Croix de la Bretonneric,
75004 Paris

Institut National de la Recherche
Agronomique (INRA)
Route de Saint-Cyr,
78000 Versailles

GABON

Directeur General de l'Energie
et des Resources Hydrauliques
B.P. 874, Libreville

FEDERAL REPUBLIC OF GERMANY

Interdisciplinary Project Group
for Appropriate Technology,
Technische Universitat
Lentzealle 86, D-1000 Berlin 33

GERMANY (cont.)

Appropriate Technology Cooperation
Society, Postfach 5180,
D-6236 Eschborn 1

Research Institute for International
Technology Cooperation,
Henricistr. 50,
5100 Aachen
Vereinsstrasse 305

DSE- German Foundation for
International Development,
Food and Agricultural Development Cntr.
Wielingerstrasse 52
D-8133 Feldafing

GHANA

Council for Scientific and
Industrial Research,
Crops Research Centre,
P.O. Box 74, Kade

Forest Products Research Institute,
University P.O. Box 63,
Kumasi

Technology Consultancy Centre (TCC)
University of Science and Technology,
Kumasi

GREECE

National Energy Council
Ministry of Co-ordination
Stadiou 3, Athens

National Energy Council
42 Akademias St.
Athens

GUATEMALA

Applied Research Division,
Central American Research
Institute for Industry (ICAITI)
P.O. Box 1552
Avenida La Reforma 4-47 Zona 10
Guatemala City

Experimental Station at Choqui
Apartado Postal 159
Quezaltenago

GUINEA

Mr. K.K. Kourouma,
Directeur des Sciences Exactes
a l'Institute Nacionale des
Recherces et Documentation,
B.P. 561, Conakry

Faculte d'Electrotechnique
I.P.G.A.N.
Conakry

Institut Polytechnique Gamal
Abdel Nasser (IPGAN)
B.P. 1147
Conakry

GUINEA BISSAU

Mr. F. Vaz Martins
Commissaire d'Etat a l'Energie,
l'Industrie et les Ressources
Naturelles,
C.P. 353, Bissau

GUYANA

Institute of Applied Science
and Technology,
University Campus,
Turkeyen, Gtr. Georgetown

National Science Research Council,
P.O. Box 689,
Georgetown.

Energy Unit
c/o Upper Mazaruni Development
Authority (Gov't of Guyana)
P.O. Box 903, Georgetown

HONDURAS

Appropriate Technology Program
Physics Department,
University of Honduras,
Tegucigalpa, D.C.

Rural Technology Programme
Apartado Postal 703,
Tegucigalpa, D.C.

Technical Cooperative Centre for
Industry, Av. La Paz 407
Apartado Postal 938,
Tegucigalpa

ICELAND

Agricultural Research Institute,
Keldnaholt,
110 Reykjavik

Geothermal Division (ORKUSTOFNUN)
United Nations University,
Grensasvegur 9,
108 Reykjavik

INDIA

Appropriate Technology Development Association,
Post Box 311, Gandhi Bhawan,
Lucknow 226001, Uttar Pradesh

Appropriate Technology Unit,
Allahabad Polytechnic,
Allahbad, Uttar Pradesh 211002

Appropriate Technology Unit,
Gov't of India, Ministry of Industry,
Dept. of Industrial Development,
268, Udyog Bhavan,
New Delhi 110011

Centre for the Application of Science and Technology to Rural Areas (ASTRA)
Indian Institute of Science,
Bangalore 560 012

Birla Institute of Technology and Sciences, Pilani,
Gobar Gas Research Stn.
Ajitmal, Etawah, Uttar Pradesh

Indian Council of Agricultural Research,
Scheme on Biogas Technology,
Sri Parasakthi College for Women,
Courtallem- 627802 Tenkasi,
Tamil Nadu

Centre for Rural Development and Appropriate Technology,
Hanz Khas, New Delhi, 110016

Jyoti Solar Energy Institute,
P.B. No.20,
Vallabah Vidyanagar 388120,
Kaira, Gujarat

INDIA (cont.)

Khadi and Village Industries Commission, Gobar Gas Scheme,
Research and Development Centre,
Kora Gramadyog, Kendra, Borivli
Bombay 400092

Resources Development Institute
E6-00
Arera Colony, P.O. Bhopal-14
462014

Shri AMM Murugappa Chettiar Research Center,
11/12 North Beach Road,
Madras 600 001

Shri AMM Murugappa Chettiar Research Centre,
Photosynthesis and Energy Div.,
Tharamani, Madras 600 042

Central Fuel Research Institute,
Dhanbad 828108, Bihar

Central Mechanical Engineering Research, Mahatma Gandhi Av.,
Durgapur 713-209

Central Power Institute,
Bangalore 560012

Energy Research Centre
Indian Institute of Technology,
Madras 600036

Indian Institute of Technology,
Chemistry Department,
Powai, Bombay 400076

International Crop Research Institute for Semi-Arid Tropics (LCRISAT)
Andhra Pradesh 802 324

Centre for Appropriate Technology,
Thirumalai, Chunkankadai P.O.
PIN 629807 Tamil Nadu

Solar Energy Laboratory,
Heat Power Engineering Dept.,
College of Engineering,
Peraringer Anna University of Technology, Guindy, Madras 600 025

INDIA (cont.)

Solar Energy Section,
Central Arid Zone Research Institute,
Jodhpur, 342003 Rajasthan

TATA Energy Research Institute,
Documentation Centre,
Bombay House, 24 Homi Mody St.,
Bombay, Maharashtra, 400 023

TATA Energy Research Institute,
Field Research Unit,
c/o Shri Aurobindo Ashran
Pondicherry 605 002

Biomass Centre,
School of Biological Sciences,
Madurai Kamaraj University
Madurai 625 021

INDONESIA

Development Technology Centre
Institute of Technology Bandung,
JI, Ganesha 10, Bandung

Solar Energy Research Center,
Gajah Mada University,
Sekip Unit III,
Yogyakarta

IRELAND

National Board for Science and
Technology,
Shelbourne House,
21 Shelbourne Road,
Dublin 4

Biomass Conversion Technical
Information Service, Institute for
Industrial Research and Standards,
(IIRS), Ballymun Road,
Dublin 9

Irish Peat Development Authority
Bord na Mona
Droichead Nua, Co. Kildare

ISRAEL

Applied Research Institute,
Ben Gurion University of the Negev,
P.O.Box 1025, Beersheva 84120

ISRAEL (cont.)

Institute of Desert Research
Sde Boker Campus, Ben Gurion
University, Sde Boker 84990

Research and Development Division,
Ministry of Energy and Infra-
structure,P.O.Box 1442,
University Str.
Tel Aviv

The Scientific Research Foundation
of Jerusalem, P.O.Box 3745,
Jerusalem 91036

Hebrew University of Jerusalem,
Dept. of Medical Ecology,
Bayit Vegan, Jerusalem 91904

Hebrew University of Jerusalem,
Dept. of Biology,
Givat Ram, Jerusalem 91904

ITALY

Centro Studi Energia,
Via Serra 6, Genova 16122

Fiat - Energy Sector,
Soft Energy Systems,
Via Cuneo 20,10153 Torino

Food and Agriculture Organization
(FAO) of the United Nations,
Via delle Termi di Caracalla,
00100 Roma

Instituto Economia delle Fonti
Di Energia,
Universita Bocconni, Milan

International School on Solar
Energy and Other Renewable
Energy Sources (SOGESTA)
P.O.Box 65, 1-61029 Urbino

JAMAICA

Renewable Energy Development Div.,
Scientific Research Council,
P.O.Box 350,
Kingston 6

JAPAN

Institute of Applied Energy
33rd Mori Building
3-8-21 Toranomon, Minato-ku,
Tokyo 105

Solar Research Laboratory
Government Research Institute
I-Chome Hirate-machi
Kita-ku, Nagoya 462

KENYA

International Council for
Research in Agroforestry (ICRAF)
P.O.Box 30677, Nairobi

Village Technology Unit
United Nations Childrens Fund (UNICEF)
P.O.Box 44145,
Nairobi

United Nations Environment
Programme (UNEP)
P.O.Box 30522,
Nairobi

Ministry of Energy
Government of Kenya
P.O.Box 30582
Nairobi

The Mazingira Institute
P.O.Box 14550
Nairobi

Environmental Liaison Centre
P.O.Box 72461
Nairobi

Karen Village Technology Unit
Karen, Nairobi

REPUBLIC OF KOREA

Biotechnology Research Dept.
Korea Institute of Science &
Technology, P.O.Box 131,
Dongdae-Mun, Seoul.

Rural Energy Research Division,
Institute of Agricultural
Engineering and Utilization,
Office of Rural Development,
Suwon

KOREA (cont.)

Appropriate Technology Studies
Centre,
Unwol-ri, Hongdong-myun,
Hongseong-gun, Chungnam 350-11

KUWAIT

Kuwait Institute for Scientific
Research (KISR)
P.O.Box 24885
Safat, Kuwait City

LIBYA

Centre for Solar Energy Studies
(CSES), P.O.Box 8004,
Tripoli

MADAGASCAR

New Energies Inc.
B.P. 382
Antananarivo

Polytechnical Institute
University of Madagascar
B.P. 1500
Antananarivo

University Delegation for the
New Energies,
University of Madagascar
Rectorat B.P. 566
Ankatso-Antananarivo

MALAYSIA

Faculty of Agricultural Engineering
University of Agriculture,
Serdang, Selangor

MALI

Lavoratoire de l'Energie Solaire
B.P. 134, Bamako

MAURITIUS

School of Industrial Technology,
University of Mauritius,
Porte Luis

MAURITIUS (cont.)

School of Industrial Technology,
University of Mauritius,
Reduit

MEXICO

Mexican Institute of Renewable
Resources,
Dr. Vertiz 724,
Mexico City

Centro de Estudios Economicos y
Sociales del Tercer Mundo,
Cor. Porfirio Diaz 50,
San Jeronimo Lidice, Mexico 20 DF

Centro de Investigacion en
Quimica Aplicada (CIQA)
Aldema Oriente No. 371,
Saltillo, Coah.

Direccion General de Aprovecha-
miento de Agnas Salinas y
Energie Solar,
Bd. Pipila 1,
Mexico 10 D.F.

NEPAL

Biogas Newsletter,
P.O.Box 1309, Kathmandu

Research Centre for Applied Science
and Technology (RECAST)
Tribhuvan University, Kirtipur,
Kathmandu

Swiss Association for Technical
Assistance (SATA)
Jawalakhel, P.O.Box 113,
Kathmandu

NETHERLANDS

Appropriate Technology Department
Eindhoven University of Technology
Paviljoen E.4, P.O.Box 513,
Eindhoven

Institute for Waste Disposal
Postbus 184,
Amersfoort

NETHERLANDS (cont.)

International Institute for Aerial
Survey and Earth Sciences (ITC)
350 Bd. 1945
Enschede

Department of Chemical Technology
Twente University of Technology
7500 AE Enschede, P.O.Box 217

NEW ZEALAND

New Zealand Energy Research and
Development Committee,
University of Auckland,
Private Bag, Aukland

Department of Scientific and
Industrial Research, (DSIR)
Charles Fergusson Building,
Bowen Street, Wellington

Liquid Fuels Trust (LFTB)
P.O.Box 17
Wellington

New Zealand Forest Service
Forest Research Institute,
Wellington

NIGER

Office de l'Energie Solaire
ONERSOL, B.P. 621, Niamey

NIGERIA

Rural Energy Systems in Nigeria
Project,
University of Ife, Ife-Ife.

International Institute of
Tropical Agriculture,
Oyo Road, PMB 5320, Ibadan

NORWAY

Development, Research and
Action Programme (DERAP)
The Chr. Michelsen Institute,
Fantoftvegen 38,
5036 Fantoft-Bergen

PAKISTAN

Appropriate Technology Development Organization,
1-B. Street 47th,
F-7/1, Islamabad

PANAMA

Instituto de Recursos Hidraulicos y Electrificacion, Direccion de Desarrollo, Departamento de Energia y Tarifas,
Apartamento 5285,
Zona 5, Panama City

PAPUA NEW GUINEA

Appropriate Technology Unit,
University of Technology,
P.O.Box 793, Lae, Papua

South Pacific Appropriate Technology Foundation (SPATF),
P.O.Box 6937, Boroko

PHILIPPINES

Maya Pig Farms,
Angono, Rizal

Dr. E. Terrado,
Non-Conventional Energy Centre,
Ministry of Energy,
Fort Bonifacio, Metro Manila

Natural Resources Ministry,
Visayas Avenue,
Diliman, Quezon City

Institute of Agricultural Engineering and Technology,
University of the Philippines at Los Banos, Laguna

Integrated Research Centre
de la Salle University,
1401 Taft Avenue,
Manila

Non-Conventional Energy Development Centre (CNED),
Don Mariano Marcos Avenue,
Diliman, Quezon City

ROMANIA

Energy Research and Modernizing Institute,
Bd. Energeticienilor 8
R-74568 Bucharest,Sector 3

Non-Conventional Energy Research and Development Programs,
National Council for Science and Technology,
Bucharest

RWANDA

Centre for the Study and Application of Energy in Rwanda,
Universite Nacional,
Boite Postal 117, Butare

SAUDI ARABIA

Solar Programs Research Institute,
University of Petroleum and Minerals, P.O.Box 144,
Dhahran International Airport
Dhahran

SENEGAL

Centre d'Etudes et de Recherche sur les Energies Renouvelables (CERER),
B.P. 476, Dakar

SIERRA LEONE

Advisory Unit for Technology Research and Development (ASTRAD),
Faculty of Engineering,
University of Sierra Leone,
Mount Aureol, Freetown

SOUTH AFRICA

Institute of Natural Resources,
P.O.Box 375,
Pietermaritzburg

SPAIN

Centro de Estudios de la Energia
c/o Agustin de Foxa
29,1 Plante
Madrid-16

SRI LANKA

Appropriate Technology Group of
Sri Lanka,
c/o CIC Ltd., 4th floor,
Hemas Building, Bristol St,
Colombo 1

Peradeniya Faculty of Engineering,
University of Sri Lanka,
Peradeniya

Shravodaya Shramadana Movement,
77 de Soysa Road, Moratuwa

SUDAN

National Council for Research,
Institute of Energy Research,
P.O.Box 2404, Khartoum

Faculty of Engineering,
Institute of Solar Energy,
University of Khartoum

SURINAM

University of Surinam
Department of Engineering,
University Complex, Building II,
Leysweg, Paramaribo

SWEDEN

National Swedish Board for Energy
Source Development,
Box 1103,
S-163 12 Spanga

University of Goteburg,
Botanical Institute,
Carl Skottsbergs Gata 22,
S-413 19 Goteburg

Swedish International Development
Authority (SIDA)
Birgir Jarls Gatan 61,
S-10525 Stockholm

Swedish Agency for Research Cooperation
with Developing Countries (SAREC)
Birgir Jarls Gatan 61
S-10525 Stockholm

SWEDEN (cont.)

Beijer Institute
Royal Swedish Academy of Sciences,
Fach S-10505,
Stockholm

IFIAS,
c/o Professor C.G. Heden
Karolinska Institute
10401 Stockholm

SWITZERLAND

International Union for the Conservation of Nature and Natural
Resources (IUCN)
Avenue du Mont Blanc,
CH-1196 Gland

Swiss Association for Appropriate
Technology (SAGAT)
Postfach 2121, CH-8028 Zurich

Swiss Association for Intermediate
Technology (SVMT)
Postfach 2134, CH-8028 Zurich

Swiss Centre for Appropriate
Technology (SKAT),
Institute of St. Gall University,
Varnbuelstr. 14,
CH-9000 St. Gallen

Project Biogas
National Energy Research Foundation
CH-8001 Zurich

TANZANIA

Arusha Appropriate Technology
Project (AATP),
S.L.P. 764, Arusha.

Faculty of Agriculture, Forestry
and Veterinary Science,
University of Dar Es Salaam,
Box 643, Motorgoro

Small Industries Development
Organization (SIDO),
P.O.Box 2476, Dar Es Salaam

THAILAND

Asian Institute of Technology (AIT),
Renewable Energy Resources,
P.O.Box 2754, Bangkok.

King Mongkuts Institute of Technology,
School of Energy and Materials,
Bangmod, Bangkok 14

Technical Division,
National Energy Administration,
Pibulthum Villa,
Kasatsuk Bridge, Bangkok 5

Thailand Institute of Scientific
and Technological Research (TISTR),
196 Phaholyothin Road,
Bangkhen, Bangkok 9.

TOGO

Direction de l'Hydraulique et de
l'Energie,
B.P. 335, Lome

Solar Energy Department,
School of Sciences,
University of Benin,
B.P. 1515, Lome

TUNISIA

Association pour le Developpement
et l'Animation Rurale (ASDEAR),
10 Rue Eve Nohelle, Tunis.

TURKEY

Turkish Solar Energy Institution,
P.O.Box 37,
Bakanliklar, Ankara

UNION OF SOVIET SOCIALIST REPUBLICS

Solar Energy Laboratory,
Academy of Sciences of USSR,
Leninski Prospekt 14,
Moscow.

All-Union Institute of Research in
Solar Technologies,
Leninski Prospekt 19,
Moscow

UNITED KINGDOM

Alternative Technology Group,
Faculty of Technology,
The Open University,
Milton Keynes, MK7 6AA

Department of Energy,
Thames House South,
Millbank,
London SW1P 4QJ

Commonwealth Forestry Institute,
Oxford University,
Oxford

National Centre for Alternative
Technology,
Llyngwern Quarry,
Machynlleth, Powys, Wales

Atomic Energy Research Establishment,
Energy Technology Support Unit,
Department of Energy,
Harwell, Didcot OX11 0RA

Commonwealth Science Council,
Marlborough House, Pall Mall,
London SW1Y 5HX

Energy Studies Unit,
University of Strathclyde,
100 Montrose Street,
Glasgow, Scotland G4 0LZ

International Institute for
Environment and Development,
10 Percy Street,
London W1P 0DR

Intermediate Technology Development Group,
9 King Street,
London WC2E 8HN

Rural Development Group,
Crown Agents, 4 Millbank,
London SW1P 3 JD

Tropical Products Institute,
56/62 Gray's Inn Road,
London, WC1X 8LU

UNITED KINGDOM (cont.)

University College,
Department of Mechanical Engineering
and Energy Studies,
Solar Energy Unit,
Newport Road,
Cardiff CF2 1TA

University of Reading,
Dept. of Engineering,
The Energy Group,
Whiteknights,
Reading, Berks. RG6 2AY

UNITED STATES OF AMERICA

Appropriate Technology International (ATI),
1709 N Street, N.W.,
Washington, D.C. 20036

The Bio-Energy Council,
Suite 825A,
1625 Eye St. N.W.,
Washington,D.C. 20006

Brookhaven National Laboratory,
Upton, Long Island, N.Y. 11973

Center for Development Technology,
Washington University,
P.O.Box 1106,
St. Louis, Missouri 63130

Cornell University Energy Programs,
Riley-Robb Hall,
Cornell University,
Ithaca, N.Y. 14853

Energy Resources Group,
University of California,
Rm 100, Bldg. T-4,
Berkeley, California 94720

The Institute for Energy Research,
Energy Management Training Program,
SUNY at Stony Brook,
Stony Brook, N.Y. 95465

National Academy of Sciences -
National Research Council,
2101 Constitution Avenue,
Washington, D.C. 20418

UNITED STATES OF AMERICA (cont.)

National Center for Appropriate
Technology.
P.O.Box 3838,
Butte, Montana 59701

National Gasohol Commission, Inc.
521 So. 14th Street, Suite 5,
Lincoln, Nebraska 68508

Office of Conservation and Solar
Energy,
U.S. Dept. of Energy, Rm. 2221C,
20 Massachusetts Ave. N.W.
Washington D.C. 20585

Peace Corps,
806 Connecticut Ave N.W.
Washington D.C. 20525

Solar Energy Research Institute
(SERI),
1617 Cole Blvd.
Golden, Colorado 80401

United Nations Development Programme (UNDP),
One United Nations Plaza,
New York, N.Y. 10017

U.S. Agency for International
Development,
Washington D.C. 20523

U.S. Department of Energy,
Division of Biological Energy
Research,
Office of Basic Energy Sciences,
Washington, D.C. 20545

University of Florida,
Solar Energy and Energy Conversion
Laboratory,
Gainesville, Florida 32611

Volunteers in Technical Assistance, Inc. (VITA)
3706 Rhode Island Ave.,
Mt. Ranier, Maryland 20822

World Bank
1818 H Street, N.W.
Washington D.C. 20433

UNITED STATES OF AMERICA (cont.)

World Watch Institute,
1776 Massachusetts Ave. N.W.
Washington D.C. 20036

UPPER VOLTA

General Office of Environment
B.P. 7044, Ouagadougou

Societe Africaine d'Etudes et de
Developpement (SAED)
B.P. 593 Ouagadougou

VENEZUELA

Ministry of Energy and Mines
Division of Non-Conventional Energy,
Centro Simon Bolivar, Caracas

Simon Bolivar University,
Chemistry Department,
Apartado 314,
La Guaira 1160-A

Readers are requested to send addresses suitable for addition to this list to P.A. Moss and D.O. Hall, King's College London, 68 Half Moon Lane, London SE24 9JF

ACKNOWLEDGEMENTS

The authors gratefully acknowledge the assistance of the following people and sources in compiling the preceeding address list:

International Directory of Appropriate Technology Resources, Volunteers in Technical Assistance (VITA), Mt. Ranier, Maryland.

OECD Directory of Applied Technology Centres, Development Centre of the Organisation for Economic Co-operation and Development, Paris.

Earthscan, 10 Percy Street, London.

Wendy Jackson and the International Directory of New and Renewable Energy Information Sources and Research Centers, SERI, Golden, Colo.

Robin Roy, Energy Research Group, The Open University, Milton Keynes.

P.K.R Nair, International Council for Research in Agroforestry, Nairobi.